A PRACTICAL GUIDE FOR STUDYING CHUA'S CIRCUITS

WORLD SCIENTIFIC SERIES ON NONLINEAR SCIENCE

Editor: Leon O. Chua
University of California, Berkeley

*To view the complete list of the published volumes in the series, please visit:
http://www.worldscibooks.com/series/wssnsa_series.shtml

WORLD SCIENTIFIC SERIES ON
NONLINEAR SCIENCE   Series A   Vol. 71

Series Editor: Leon O. Chua

A PRACTICAL GUIDE FOR STUDYING CHUA'S CIRCUITS

Recai Kılıç

Erciyes University, Turkey

World Scientific

NEW JERSEY · LONDON · SINGAPORE · BEIJING · SHANGHAI · HONG KONG · TAIPEI · CHENNAI

Published by

World Scientific Publishing Co. Pte. Ltd.

5 Toh Tuck Link, Singapore 596224

USA office: 27 Warren Street, Suite 401-402, Hackensack, NJ 07601

UK office: 57 Shelton Street, Covent Garden, London WC2H 9HE

British Library Cataloguing-in-Publication Data
A catalogue record for this book is available from the British Library.

A PRACTICAL GUIDE FOR STUDYING CHUA'S CIRCUITS
World Scientific Series on Nonlinear Science, Series — Vol. 71

ISBN-13 978-981-4291-13-2
ISBN-10 981-4291-13-7

Printed in Singapore.

Dedicated to my parents.

Preface

Many chaotic circuit models have been developed and studied up to date. Autonomous and nonautonomous Chua's circuits hold a special importance in the studies of chaotic system modeling and chaos-based science and engineering applications. Since a considerable number of hardware and software-based design and implementation approaches can be applied to Chua's circuits, these circuits also constitute excellent educative models that have pedagogical value in the study of nonlinear dynamics and chaos.

In this book, we aim to present some hardware and software-based design and implementation approaches on Chua's circuits with interesting application domain examples by collecting and reworking our previously published works. The book also provides new educational insights for practicing chaotic dynamics in a systematic way in science and engineering undergraduate and graduate education programs. We hope that this book will be a useful practical guide for readers ranging from graduate and advanced undergraduate science and engineering students to nonlinear scientists, electronic engineers, physicists, and chaos researchers.

Organization of the book

Chapter 1 is devoted to autonomous Chua's Circuit which is accepted as a paradigm in nonlinear science. After comparing the circuit topologies proposed for Chua's circuit, the chapter presents several alternative hybrid realizations of Chua's circuit combining circuit topologies proposed for the nonlinear resistor and the inductor element in the literature.

Numerical simulation and mathematical modeling of a linear or nonlinear dynamic system plays a very important role in analyzing the system and predetermining design parameters prior to its physical realization. Several numerical simulation tools have been used for simulating and modeling of nonlinear dynamical systems. In that context, Chapter 2 presents the use of MATLAB^{TM} and SIMULINK^{TM} in dynamic modeling and simulation of Chua's circuit.

Field programmable analog array (FPAA) is a programmable device for analog circuit design and it can be effectively used for programmable and reconfigurable implementations of Chua's circuit. FPAA is more efficient, simpler and economical than using individual op-amps, comparators, analog multipliers and other discrete components used for implementing Chua's circuit and its changeable nonlinear structure. By using this approach, it is possible to obtain a fully programmable Chua's circuit which allows the modification of circuit parameters on the fly. Moreover, nonlinear function blocks used in this chaotic system can be modeled with FPAA programming and a model can be rapidly changed for realizing another nonlinear function. In Chapter 3, we introduce FPAA-based Chua's circuit models using different nonlinear functions in a programmable and reconfigurable form.

In Chapter 4, we describe an interesting switched chaotic circuit using autonomous and nonautonomous Chua's circuits. It is called as "Mixed-mode chaotic circuit (MMCC)". After introducing the original design of MMCC, alternative circuit implementations of the proposed circuit are given in the Chapter.

In order to operate in higher dimensional form of autonomous and nonautonomous Chua's circuits while keeping their original chaotic behaviors, we modified the voltage mode operational amplifier (VOA)-based autonomous Chua's circuit and nonautonomous Murali-Lakshmanan-Chua (MLC) circuit by using a simple experimental method. In Chapter 5, this experimental method and its application to autonomous and nonautonomous Chua's circuits are introduced with simulation and experimental results.

In Chapter 6, we discuss some interesting synchronization applications of Chua's circuits. Besides Chua's circuit realizations described in the previous chapters, some synchronization applications of state-controlled cellular neural network (SC-CNN)-based circuit which is a different version of Chua's circuit are also presented in the Chapter.

In Chapter 7, a versatile laboratory training board for studying Chua's circuits is introduced with sample laboratory experiments. The issues presented in this chapter are for education purposes and they will contribute to studies on nonlinear dynamics and chaos in different disciplines.

Acknowledgements

I would like to thank the following colleagues who contributed to my study, and the editing process of the book:

Prof. Dr. Mustafa ALÇI	Erciyes University
Prof. Dr. Hakan KUNTMAN	İstanbul Technical University
Prof. Dr. Uğur ÇAM	Dokuz Eylül University
Dr. Enis GÜNAY	Erciyes University
Dr. Esma UZUNHİSARCIKLI	Erciyes University
Dr. Muzaffer Kanaan	Erciyes University
Research Assist. Fatma Y.DALKIRAN	Erciyes University
Researcher Barış KARAUZ	HES Company

I would like to state my special thanks to my doctoral advisor, Prof. Dr. Mustafa ALÇI for encouraging me to study chaotic circuits and systems during my graduate program.

I would also like to thank Prof. Leon Chua for his encouragement and recommendation to publish this book in the World Scientific Nonlinear Science, Series A.

Recai Kılıç
Kayseri, Turkey, November 2009

Contents

Chapter 1

Autonomous Chua's Circuit: Classical and New Design Aspects

In this chapter, we will focus on the autonomous Chua's circuit [24], which is shown in Fig. 1.1 containing three energy storage elements, a linear resistor and a nonlinear resistor N_R, and its discrete circuitry design and implementations. Since Chua's circuit is an extremely simple system, and yet it exhibits a rich variety of bifurcations and chaos among the chaos-producing mechanisms [see for example 2, 9, 18, 24, 34–35, 39, 45, 52, 58, 82, 87, 89, 101–104, 109, 119, 121, 131–132, 135, 140–141, 146, 149, and references therein], it has a special significance. The aim of this chapter is to show that several hardware design techniques can be adapted to this paradigmatic circuit, and alternative experimental setups can be constituted by using different Chua's circuit configurations for practical chaos studies.

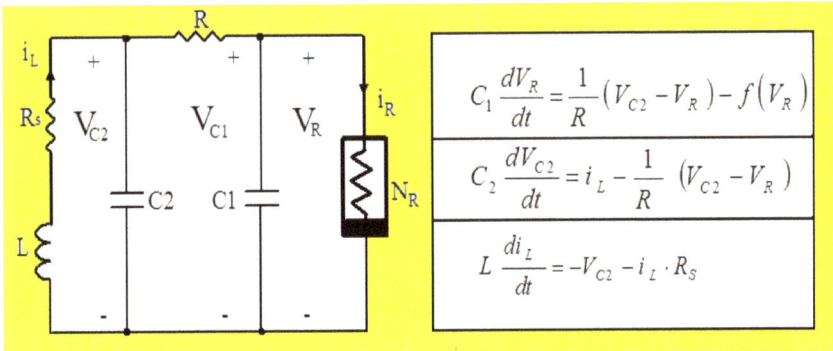

$$C_1 \frac{dV_R}{dt} = \frac{1}{R}(V_{C2} - V_R) - f(V_R)$$

$$C_2 \frac{dV_{C2}}{dt} = i_L - \frac{1}{R}(V_{C2} - V_R)$$

$$L \frac{di_L}{dt} = -V_{C2} - i_L \cdot R_S$$

Fig. 1.1 Autonomous Chua's circuit.

Several realizations of Chua's circuit have been proposed in the literature. The methodologies used in these realizations can be divided into two basic categories. In the first approach, a variety of circuit topologies have been considered for realizing the nonlinear resistor N_R in Chua's circuit. The main idea in the second approach related to the implementation of Chua's circuit is an inductorless realization of Chua's circuit. In this chapter, after comparing the circuit topologies proposed for Chua's circuit, several alternative hybrid realizations of Chua's circuit combining circuit topologies proposed for the nonlinear resistor and the inductor element in the literature are presented.

1.1 The Nonlinear Resistor Concept and Chua's Diode

The term *Chua's diode* is a general description for a two-terminal nonlinear resistor with a piecewise-linear characteristic. In the literature, Chua's diode is defined in two forms [53]. As shown in Fig. 1.2(a), the first type of Chua's diode is a voltage-controlled nonlinear element characterized by $i_R = f(v_R)$, and the other type is a current-controlled nonlinear element characterized by $v_R = g(i_R)$.

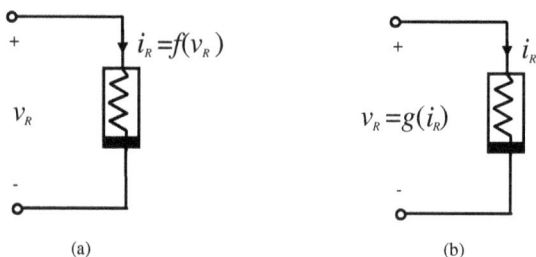

Fig. 1.2 (a) Voltage-controlled Chua's diode, (b) current-controlled Chua's diode.

Chaotic oscillators designed with Chua's diode are generally based on a single, three-segment, odd-symmetric, voltage-controlled piecewise-linear nonlinear resistor structure. Such a voltage-controlled characteristic of Chua's diode is given in Fig. 1.3.

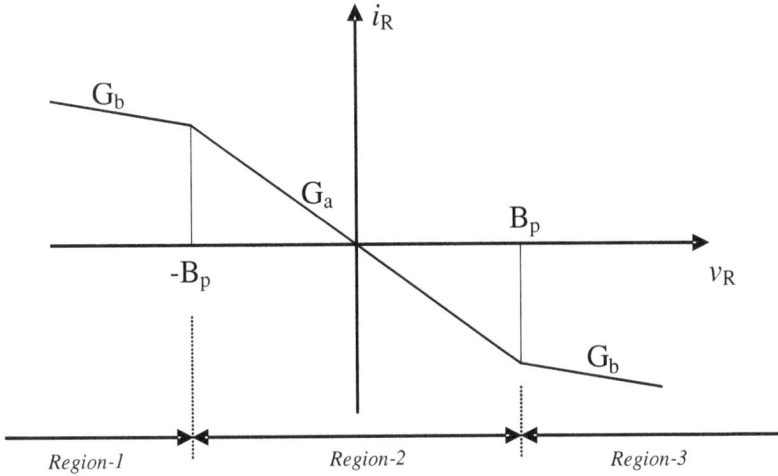

Fig. 1.3 Three-segment odd-symmetric voltage-controlled piecewise-linear characteristic of Chua's diode.

This characteristic is defined by

$$i_R = f(v_R) = G_b v_R + 0.5(G_a - G_b)\left[\left|v_R + B_p\right| - \left|v_R - B_p\right|\right]$$

$$= \begin{cases} G_b v_R + (G_b - G_a)B_p, & v_R \langle -B_p \\ G_a v_R, & -B_p \leq v_R \leq B_p \\ G_b v_R + (G_a - G_b)B_p, & v_R \rangle B_p \end{cases} \quad (1.1)$$

In this definition, G_a and G_b are the inner and outer slopes, respectively, and $\pm B_p$ denote the breakpoints. Now, let us demonstrate why Eq. (1.1) defines the $(v - i)$ characteristic of Fig. 1.3. For this Piecewise-Linear (PWL) analysis, our starting point is the "concave resistor" concept [22]. The concave resistor is a piecewise-linear voltage-controlled resistor uniquely specified by (G, B_p) parameters. Symbol, characteristic and equivalent circuit of the concave resistor is shown in Fig. 1.4. The functional representation of the concave resistor is given as follows:

$$i = \frac{1}{2}G\left[\left|v - B_p\right| + (v - B_p)\right] \quad (1.2)$$

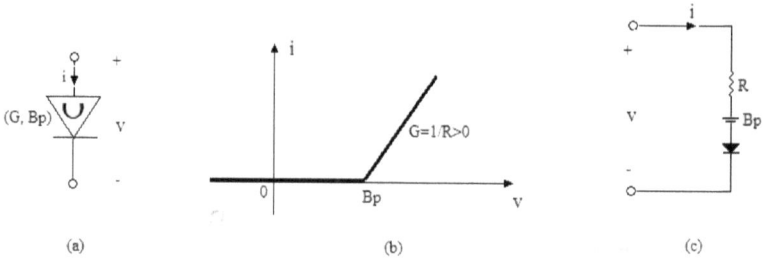

Fig. 1.4 For a concave resistor, (a) symbol, (b) characteristic and (c) equivalent circuit.

This representation can be proved by adding the plot of the term $i = (G/2)(v\text{-}B_p)$ and its absolute value term $i = (G/2)|v\text{-}B_p|$ as shown in Fig. 1.5. Now, let us consider the piecewise-linear characteristic of Chua's diode in Fig. 1.3. The three linear segments have slopes as shown in the figure:

$$\text{Region } 1: G = G_b$$
$$\text{Region } 2: G = G_a \qquad\qquad (1.3)$$
$$\text{Region } 3: G = G_b$$

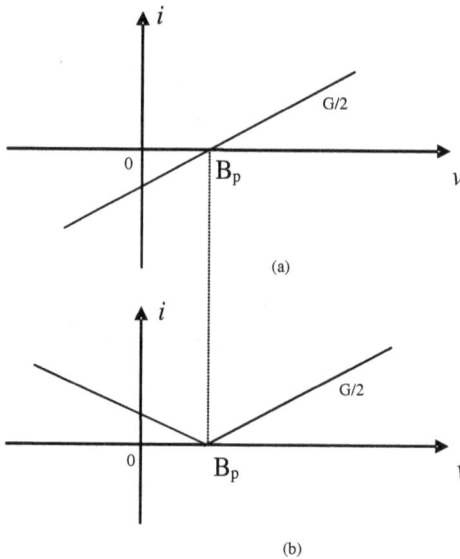

Fig. 1.5 Graphical illustration of Eq. (1.2), (a) $i = (G/2)(v - B_p)$, (b) $i = (G/2)|v - B_p|$.

(a)

(b)

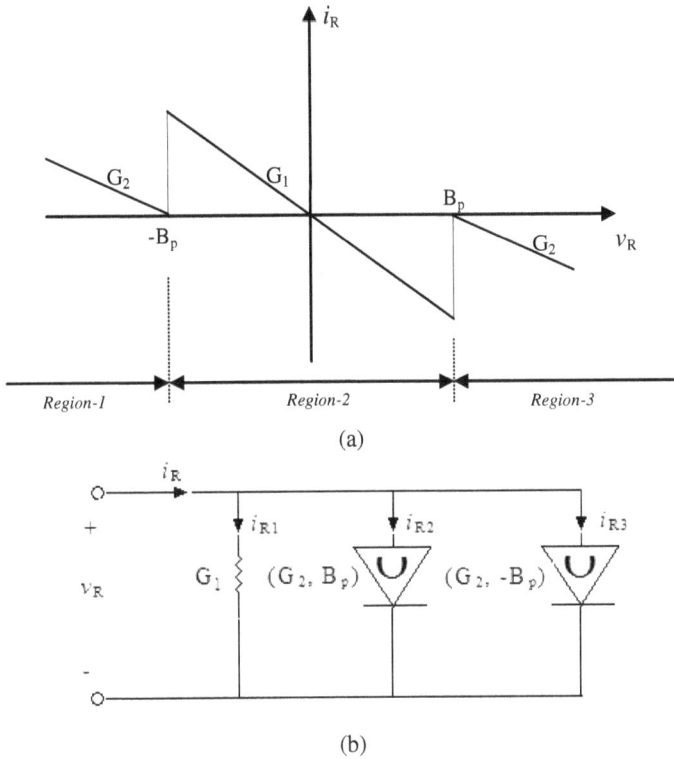

Fig. 1.6 (a) Decomposition of the characteristic in Fig. 1.3, (b) equivalent circuit for the decomposed characteristic in Fig. 1.6(a).

This characteristic can be decomposed into a sum of three components as shown in Fig. 1.6(a). These components are a straight line passing through the origin with slope G_1, a concave resistor with a negative slope G_2 and $(+B_p)$ breakpoint, and a concave resistor with the same negative slope G_2 and $(-B_p)$ breakpoint. The corresponding circuit is shown in Fig. 1.6 (b). The driving-point characteristic of the circuit can be obtained by adding three branch currents:

$$i_R = i_{R1} + i_{R2} + i_{R3} = \hat{i}(v_R)$$
(1.4)

The characteristic of the inner region is defined as

$$i_{R1} = G_1 v_R$$
(1.5)

and the characteristic of the first concave resistor with positive breakpoint is given by

$$i_{R2} = \frac{1}{2}G_2\left[\left|v_R - B_p\right| + \left(v_R - B_p\right)\right] \tag{1.6}$$

Because the second concave resistor's characteristic is symmetrical to the first concave resistor's characteristic, we should use the $i_R = -f(-v_R)$ function in Eq.(1.6) to obtain the characteristic of the second concave resistor. In this case, the characteristic of the second concave resistor is stated by

$$i_{R3} = -\frac{1}{2}G_2\left[\left|-v_R - B_p\right| + \left(-v_R - B_p\right)\right] \tag{1.7}$$

Combining three branch currents (i_{R1}, i_{R2}, i_{R3}), we obtain

$$i_R = G_1 v_R + \frac{1}{2}G_2\left[\left|v_R - B_p\right| + \left(v_R - B_p\right)\right] + \frac{1}{2}G_2\left[-\left|v_R + B_p\right| + \left(v_R + B_p\right)\right] \tag{1.8}$$

To make Eq. (1.8) identical with the PWL characteristic of Chua's diode, the parameters (G_1, G_2) and (G_a, G_b) must be matched as follows:

$$G_1 = G_a$$
$$G_1 + G_2 = G_b \Rightarrow G_2 = G_b - G_a \tag{1.9}$$

By using these statements in Eq. (1.8), the general form of Chua's diode is obtained as follows:

$$i_R = G_b v_R + \frac{1}{2}(G_a - G_b)\left[\left|v_R + B_p\right| - \left|v_R - B_p\right|\right] \tag{1.10}$$

1.2 Circuit Topologies for Realization of Chua's Diode

This section discusses several circuitry designs of Chua's diode. After giving various circuit realizations for Chua's diode, we compare these realizations with respect to circuit design issues.

Several implementations of Chua's diode already exist in the literature. Early implementations use diodes [94], op amps [92, 57], transistors [93] and OTAs [29]. One of the earliest implementations of

Chua's diode implemented by Matsumoto *et al.* [94] is shown in Fig. 1.7(a).

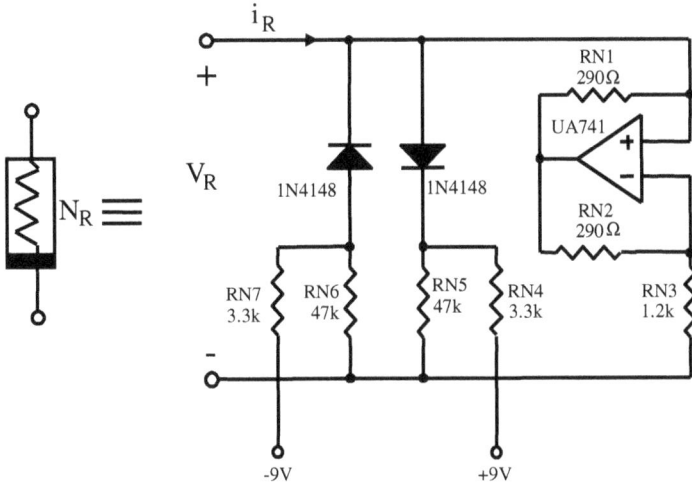

(a)

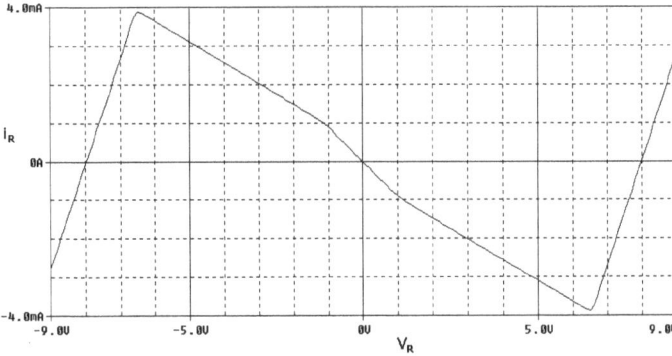

(b)

Fig. 1.7 (a) The circuit structure of Chua's diode implemented by Matsumoto *et al.* [94], (b) simulated v-i characteristic of Chua's diode of Fig. 1.7(a).

As shown in the figure, Chua's diode is realized by means of an op amp with a pair of diodes, seven resistors and DC supply voltages of $\pm 9V$, yielding $G_a \cong -0.8$ mA/V, $G_b \cong -0.5$ mA/V and $\pm B_p = \pm 1$ V. The simulated v-i characteristic of Chua's diode of Fig. 1.7(a) is shown in

Fig. 1.7(b). Cruz & Chua [29] designed the first monolithic implementation of Chua's diode using the OTA-based circuit topology in Fig. 1.8(a).

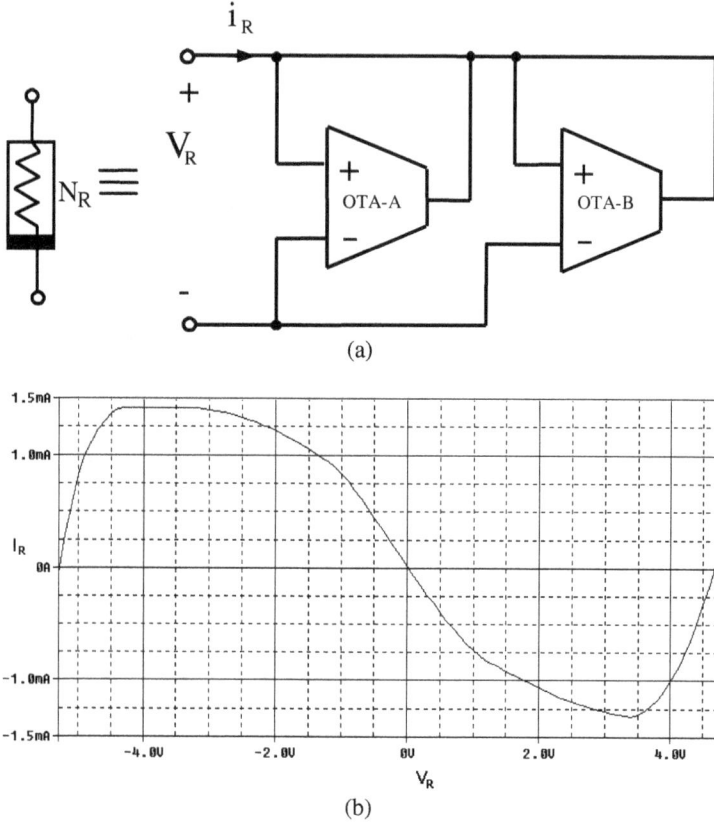

(a)

(b)

Fig. 1.8 (a) The OTA-based nonlinear resistor structure proposed by Cruz & Chua [29], (b) Simulated v-i characteristic of Chua's diode of Fig. 1.8(a).

As indicated in Fig. 1.8(a), this realization is based on only two OTAs, and the dc parameters are determined as $G_a \cong -0.78$ mA/V, $G_b \cong -0.41$ mA/V and $\pm B_p = \pm 0.7$ V. The simulated v-i characteristic of Chua's diode of Fig. 1.8(a) is shown in Fig. 1.8(b). Other sample monolithic implementations of Chua's diode have been reported [120, 122].

A realization of Chua's diode proposed by Kennedy [57], which is designed by connecting two voltage-controlled negative impedance converters in parallel, has been accepted as the standard for discrete implementation. This op amp–based nonlinear resistor structure with its simulated dc characteristic is shown in Fig. 1.9.

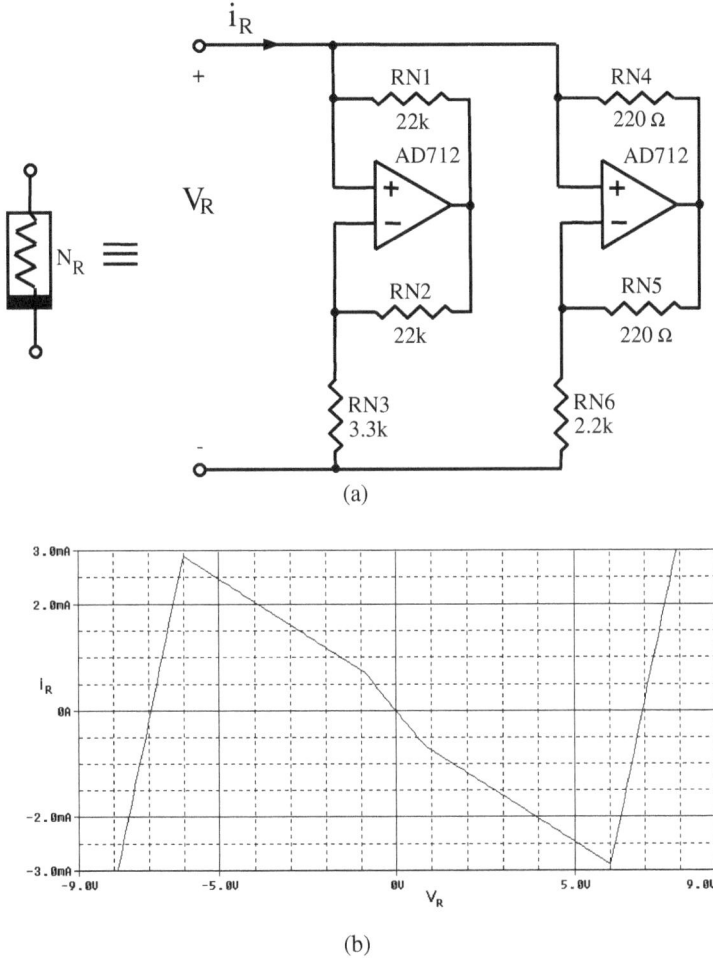

(a)

(b)

Fig. 1.9   (a) The op amp–based nonlinear resistor structure implemented by Kennedy [57], (b) simulated v-i characteristic of the nonlinear resistor of Fig. 1.9(a).

As shown in Fig. 1.9(a), this realization uses two op amps, operating in both their linear and nonlinear region, and six resistors. The slopes and breakpoints are chosen as $G_a \cong -0.756$ mA/V, $G_b \cong -0.409$ mA/V and

$\pm B_p = \pm 1$ V with the circuit parameters in Fig. 1.9(a). As Chua's circuit has a simple and easily configurable circuit structure, most of the experimental studies with it in the literature have been performed using this standard VOA-based implementation.

Due to the frequency limitations of the voltage op amps, VOA-based Chua's diode implementations impose an upper limit on the operating frequency. Therefore, in the literature new design ideas for implementing Chua's diode are considered aiming for high-frequency chaotic signals. Two alternative implementations of a VOA-based Chua's diode have been presented by Senani & Gupta [128] and Elwakil & Kennedy [33]. The proposed nonlinear resistor circuit topologies are shown in Fig. 1.10 and Fig. 1.11, respectively.

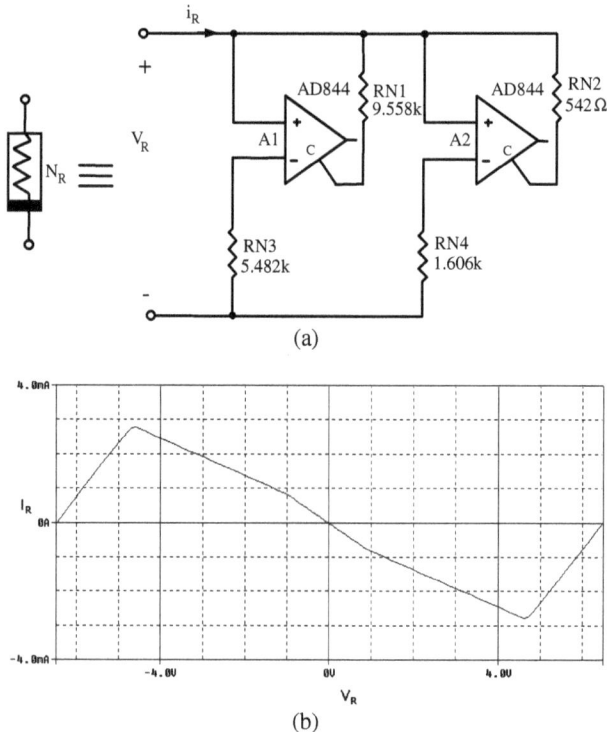

(a)

(b)

Fig. 1.10 (a) The CFOA-based nonlinear resistor structure proposed by Senani & Gupta [128], (b) simulated v-i characteristic of the nonlinear resistor of Fig. 1.10(a).

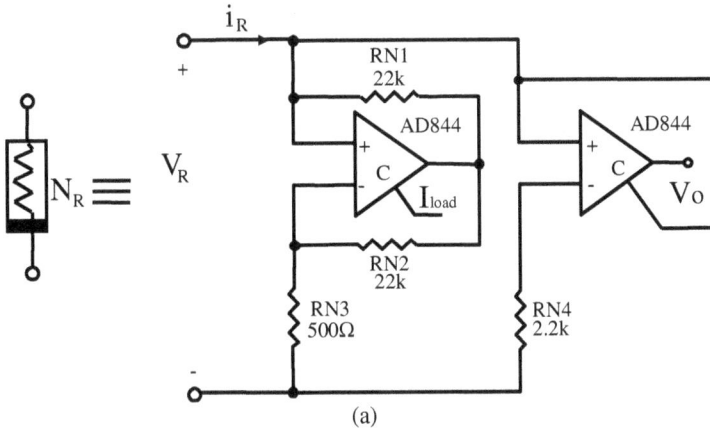

(a)

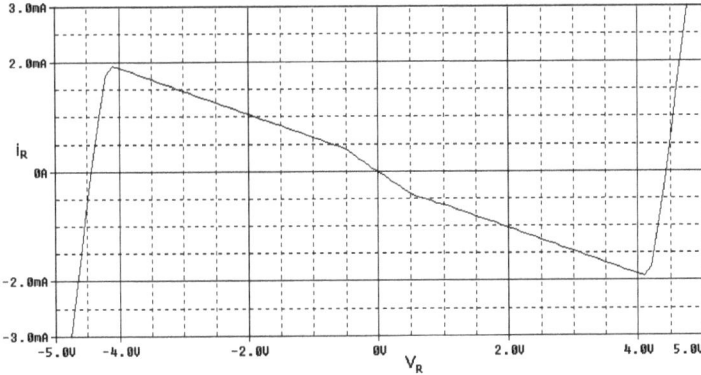

(b)

Fig. 1.11 (a) The CFOA-based nonlinear resistor structure proposed by Elwakil & Kennedy [33], (b) simulated v-i characteristic of the nonlinear resistor of Fig. 1.11(a).

In these implementations, the authors aim to use the voltage-current capabilities of a current feedback op amp (CFOA), which offers several advantages over a classic voltage op amp. In the circuit topology shown in Fig. 1.10(a), each of the CFOAs is configured as a negative impedance converter with resistors R_{N1} and R_{N2} shorted. In this case the circuit is basically a parallel connection of two negative resistors ($-R_{N3}$) and ($-R_{N4}$). Adding resistors R_{N1} and R_{N2} and using different power supply voltages for the two CFOAs, the authors offer the circuit realization for

Chua's diode with the following parameters: $R_{N1} = 9.558k\Omega$, $R_{N2} = 542\Omega$, $R_{N3} = 5.482k\Omega$, $R_{N4} = 1.606k\Omega$; and the power supplies $\pm V_1 = \pm 4.05$ V, $\pm V_2 = \pm 11.23$ V for A1 and A2, respectively. The use of these parameters yields $G_a \cong -0.8$ mA/V, $G_b \cong -0.5$ mA/V and $\pm B_p = \pm 1$ V. The simulated v-i characteristic of Chua's diode of Fig. 1.10(a) is shown in Fig. 1.10(b).

In another CFOA-based implementation of Chua's diode, shown in Fig. 1.11 with its simulated dc characteristic, the design methodology is similar to that of Kennedy [57]. While the authors used a CFOA connected as a CCII (second generation current conveyor) with resistor R_{N4} to operate as a linear negative impedance converter (NIC) operating primarily in its linear region in Kennedy's design, they configured resistors R_{N1}, R_{N2}, R_{N3} and the associated CFOA to operate as a nonlinear voltage-controlled negative impedance converter (VNIC) instead of the VOA and its three resistors in Kennedy's implementation. While the authors used the same values of R_{N1}, R_{N2}, and R_{N4} as in the Kennedy design [57], they determined R_{N3} by adjusting its value to observe chaos. Two AD844-type CFOAs biased with ± 9 V are used in this implementation. With these parameters, the desired dc characteristics, $G_a \cong -0.8$ mA/V and $G_b \cong -0.5$ mA/V, has been achieved.

Both of the CFOA-based designs for Chua's diode employ fewer resistors than that used in Kennedy's original VOA-based design. The circuit design in Fig. 1.11 provides a buffered output voltage that directly represents a state variable. Since one of the two output voltages can be used as the carrier signal in chaotic communication systems, the feature of a buffered and isolated voltage output directly representing a state variable in the Chua's diode design of Elwakil & Kennedy constitutes an advantage over other CFOA-based designs. Both CFOA-based Chua's diode circuit topologies can be configured such that the chaotic spectrum is extended to higher frequencies than with VOA-based implementations.

Another realization of Chua's diode [75] is shown in Fig. 1.12. This realization is based on four-terminal floating nullor (FTFN) circuit topology. The FTFN has been receiving considerable attention recently, as it has been shown that it is a very flexible and versatile building block in active network synthesis [47]. This leads to growing attention in design of amplifiers, gyrators, inductance simulators, oscillators and filters that use FTFN as an active element [14].

The nullor model and circuit symbol of FTFN are illustrated in Fig. 1.13(a) and 1.13(b), respectively.

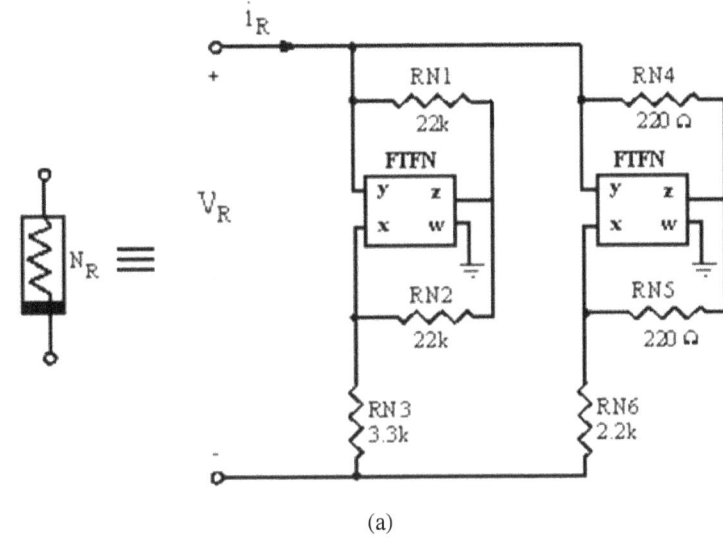

(a)

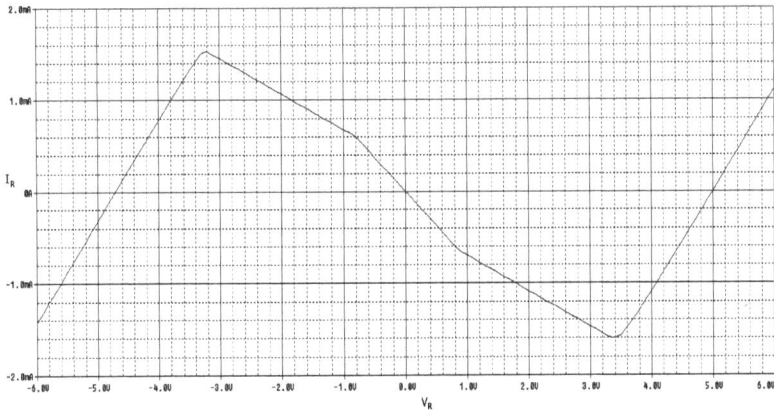

(b)

Fig. 1.12 (a) FTFN-based nonlinear resistor structure proposed by Kılıç *et al.* [75], (b) Simulated v-i characteristic of the nonlinear resistor of Fig. 1.12(a).

(a)

(b)

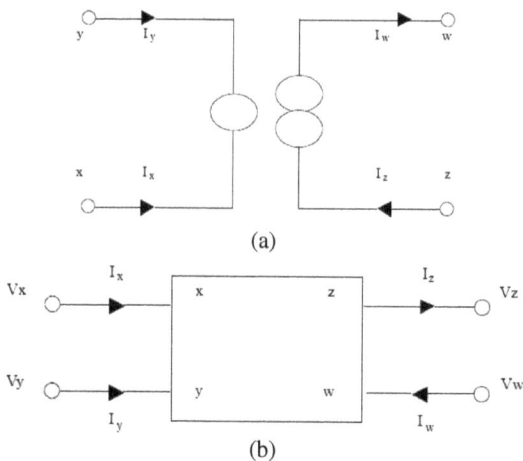

Fig. 1.13 (a) Nullor model and (b) circuit symbol of an FTFN.

The port characteristics of an FTFN can be described as

$$V_x = V_y$$
$$I_x = I_y = 0 \qquad\qquad (1.11)$$
$$I_z = I_w$$

FTFN has infinite input impedance at x and y terminals, as in a VOA. Also, FTFN has arbitrary output impedance at z and w terminals. While two voltages between the input terminals of FTFN are equivalent as in input terminals of a VOA, two currents of the floating output ports of FTFN are equivalent. Since many building blocks are represented using the same nullor and mirror elements but with different connections, it is possible to generate ideally equivalent circuits by simply replacing one building block in any circuit with another one. In the literature, several ideally equivalent building block configurations have been offered [10]. One of these realizations is that between a VOA and an FTFN. In the FTFN-based realization of Chua's diode in Fig. 1.12, by using this transformation between VOA and FTFN building blocks, two FTFN blocks instead of two VOAs in the VOA-based realization of Chua's diode in Fig. 1.9(a) can be used without any changes in other circuit elements and circuit connections.

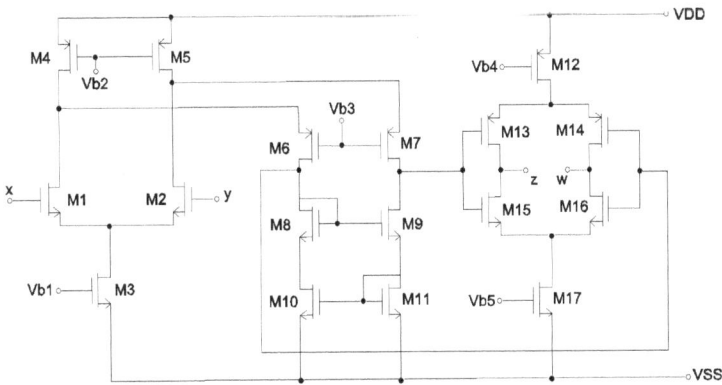

(a)

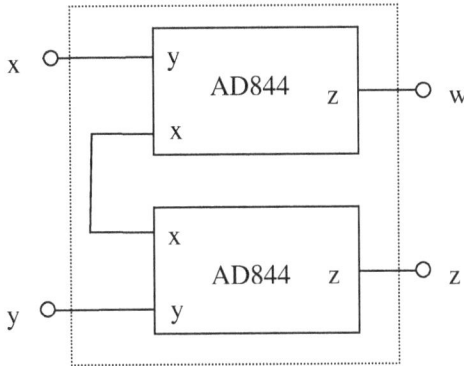

(b)

Fig. 1.14 (a) A CMOS realization of FTFN [14], (b) AD844-type current conveyor-based realization of FTFN.

Although FTFN is not commercially available, different realizations including CMOS designs for the FTFN have been suggested in the literature [14]. Also there is a practical realization that is formed by two AD844-type current conveyors. Fig. 1.14 describes these realizations for an FTFN. The simulated dc characteristic of an FTFN-based Chua's diode was obtained by using a CMOS realization of FTFN in Fig. 1.14(a).

The original chaotic behavior of Chua's circuit has also been captured with a smooth cubic nonlinearity [150], piecewise-quadratic function

[138] and some trigonometric functions [19]. The realizations of Chua's diode with these nonlinearities require a significant amount of circuitry including op amp, analog multiplier, trigonometric function generator IC. But a realization of Chua's diode with cubic-like nonlinearity by O'Donoghue *et al.* [107] offers a very simple circuit realization that consists of just four MOS transistors. This realization is shown in Fig. 1.15(a), and its cubic-like nonlinearity is defined by the following i-v characteristic.

$$i_R = f(v_R) = G_a v_R - \frac{G_a}{E_{sat}} v_R^3 \qquad (1.12)$$

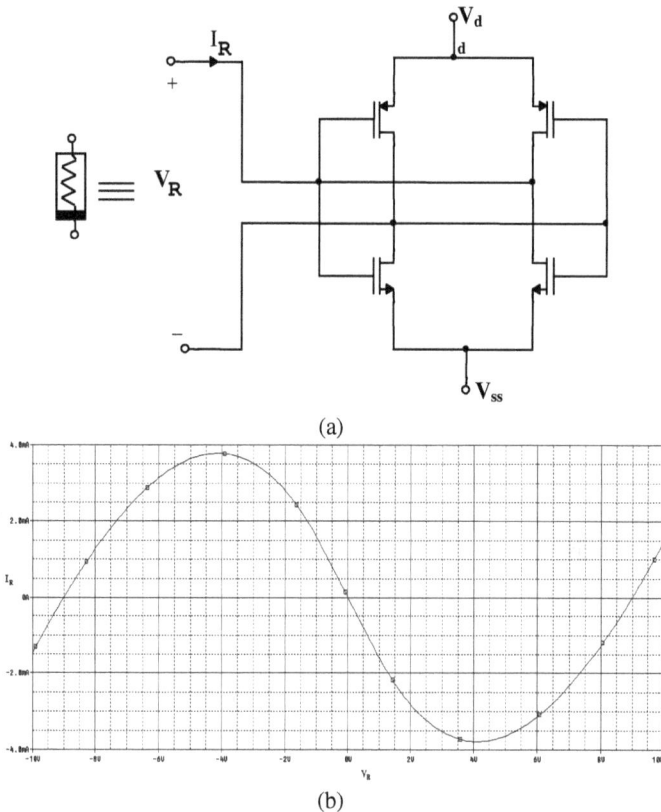

(a)

(b)

Fig. 1.15 (a) MOS transistor-based nonlinear resistor structure proposed by O'Donoghue *et al.* [107], (b) simulated v-i characteristic of the nonlinear resistor of Fig. 1.15(a).

Figure 1.15(b) shows the simulated i-v characteristic of the cubic-like Chua's diode shown in Fig. 1.15(a). In simulation studies, the following SPICE Level 3 transistor models have been used [107]:

model nmos level = 3, L = 10 u, W = 35.4 u, VTO = 1 v, lambda = 0, kp = 111 u

model pmos level = 3, L = 10 u, W = 70.8 u, VTO = –1 v, lambda = 0, kp = 55.5 u

A standard 4007 inverter IC package can be used for practical realization of the four-transistor Chua's diode. Chua's oscillator can be easily implemented by using this Chua's diode, and the resulting Chua's oscillator can operate in the MHz frequency range.

1.3 Circuit Topologies for an Inductorless Chua's Circuit

An inductorless Chua's circuit can be produced by using a synthetic inductor, *i.e.*, an inductance simulator, instead of inductor element, and by using RC configurations instead of the LC resonator in Chua's circuit. Inductorless realizations of Chua's circuit using OTA-based inductance simulators have been reported [29, 122]. In addition to OTA-based realizations, op amp–based inductance simulator structures can be used in designing inductorless Chua's circuits and other chaotic circuits [139]. Such an op amp–based inductance simulator design is shown in Fig. 1.16.

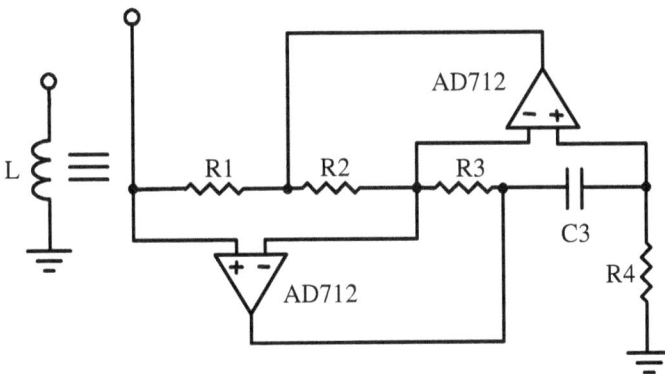

Fig. 1.16 Op amp–based inductance simulator.

The equivalent inductance value can be computed as follows:

$$L_{eq} = \frac{R_1 R_3 R_4 C_3}{R_2} \tag{1.13}$$

Due to the nonidealities of the op amps, this approach has a limited frequency range. Therefore, in experimental studies with such op amp–based inductance simulators, the nonidealities and parasitics should be taken into consideration when considering the inductor. An additional drawback is that the op amp–based inductor simulator can only be used for grounded inductance as in Chua's circuit. For simulating a floating inductance in any chaotic circuit, alternative topologies should be used. In the literature, some CFOA-based synthetic inductor structures have been used for inductorless realizations of Chua's circuit [74, 128]. These synthetic inductor structures are shown in Fig. 1.17.

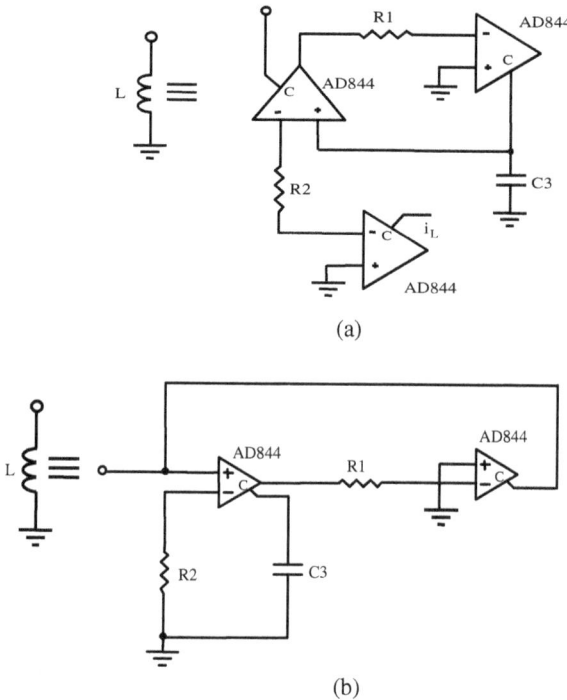

(a)

(b)

Fig. 1.17 CFOA-based synthetic inductor structures.

In these inductance simulators, the equivalent inductance value can be computed as follows:

$$L_{eq} = C_3 R_1 R_2 \qquad (1.14)$$

In the literature, an alternative inductance simulator topology based on an FTFN has been introduced for inductorless realization of Chua's circuit and other chaotic oscillators that contains both grounded and floating inductor elements [74]. The FTFN-based inductance simulator, shown in Fig. 1.18, allows one to simulate not only a grounded inductor but also a floating inductor. Although the FTFN-based inductance simulator structure is shown in floating inductor form in Fig. 1.18, this simulator may also be used as a grounded inductor by connecting one port of the floating inductance to ground. Routine analysis yields the equivalent inductance between the two terminals as

$$L_{eq} = \frac{C_3 R_1 R_2 R_3}{R_4} \qquad (1.15)$$

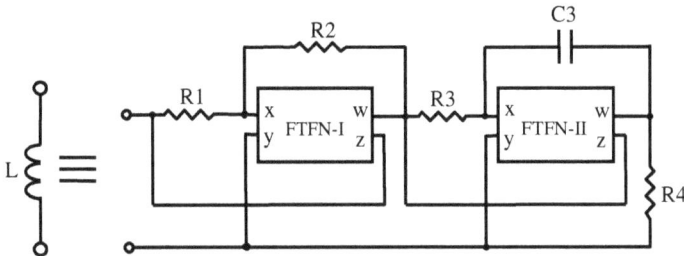

Fig. 1.18 FTFN-based inductance simulator.

In addition to using the inductance simulator for an inductorless Chua's circuit, an alternative approach has been developed based on replacing the LC resonator of Chua's circuit with an RC configuration. Such a realization has been proposed by Morgül [97]. In Morgül's implementation, shown in Fig. 1.19, the LC resonator of Chua's circuit was replaced by a Wien bridge–based circuit topology. Morgül [97] showed that with appropriate element values, the Wien bridge–based circuit topology realizes Chua's circuit.

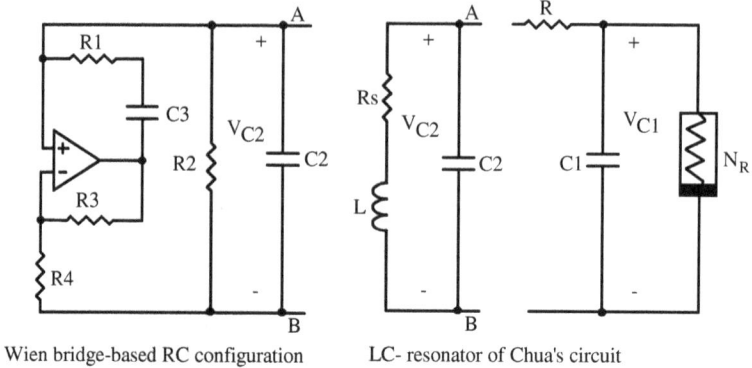

Wien bridge-based RC configuration   LC- resonator of Chua's circuit

Fig. 1.19 Wien bridge-based circuit topology, proposed by Morgül [97], for replacing with LC resonator of Chua's circuit.

Two modes (passive and oscillatory) have been identified in this implementation. In the passive mode, R, C_1 and the nonlinear resistor are kept the same in both circuits, and the input impedances of the Wien bridge part and the LC part in Fig. 1.19 are matched. In the oscillatory mode, the Wien bridge is first tuned to observe oscillations (*i.e.,* when R is open-circuited), and then by tuning R it is possible to observe chaotic oscillations for certain parameter values.

1.4 Alternative Hybrid Realizations of Chua's Circuit

In this section, we present seven hybrid realizations of Chua's circuit that exploit the circuit topologies described in the former sections [59]. Our aim is to provide several alternative realizations of Chua's circuit. The circuit structure and circuit parameters that yield a double-scroll attractor are detailed in the following subsections. First, the chaotic behaviors of the hybrid realizations are investigated by PSPICE^{TM} [118] simulation experiments. Then we show a sample experimental realization of inductorless Chua's circuit design.

1.4.1 *Hybrid-I realization of Chua's circuit*

Fig. 1.20 shows a Hybrid-I realization of Chua's circuit using two AD712 BiFET op amp biased with ± 9 V, six resistors to implement VOA-based nonlinear resistor, three AD844-type CFOA biased with ± 9 V, two resistors, and a capacitor to implement CFOA-based grounded synthetic inductor for L = 18 mH inductance value. While we used the typical parameter values for the nonlinear resistor as shown in Fig. 1.20, we configured the parameters of CFOA-based synthetic inductor as $R_1 = R_2 = 1$ kΩ, and $C_3 = 18$ nF to obtain L = 18 mH according to Eq. (1.14). The rest of the parameters are chosen as $C_1 = 10$ nF, $C_2 = 100$ nF, and R = 1700 Ω such that the circuit exhibits double-scroll attractor behavior.

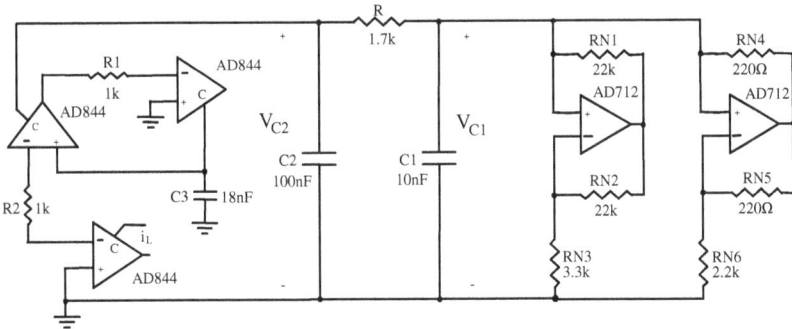

Fig. 1.20 Hybrid-I realization of Chua's circuit.

The PSPICE simulation results for the Hybrid-I realization of Chua's circuit are shown in Fig. 1.21. While the chaotic dynamics V_{C1}, V_{C2} and i_L are shown in Fig. 1.21(a), the double-scroll attractor is observed in Fig. 1.21(b). This realization uses both VOAs and CFOAs. Due to the use of CFOAs for the synthetic inductor, not only the state variables V_{C1} and V_{C2}, but also the third state variable, i_L, are accessible in a direct manner.

1.4.2 *Hybrid-II realization of Chua's circuit*

Fig. 1.22 shows a Hybrid-II realization of Chua's circuit using two AD844-type CFOAs biased with ± 9 V, four resistors to implement a CFOA-based nonlinear resistor, three AD844-type CFOAs biased with

±9 V, two resistors and a capacitor to implement the CFOA-based synthetic inductor with L = 18 mH.

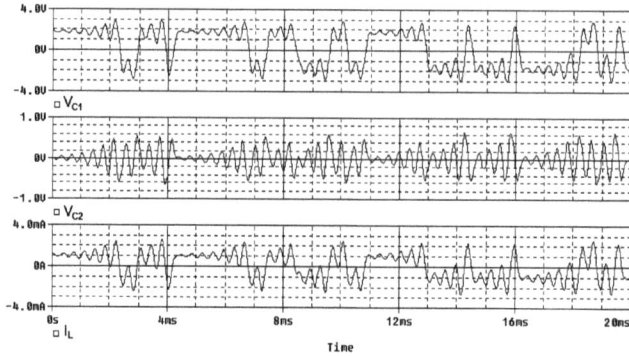

(a)

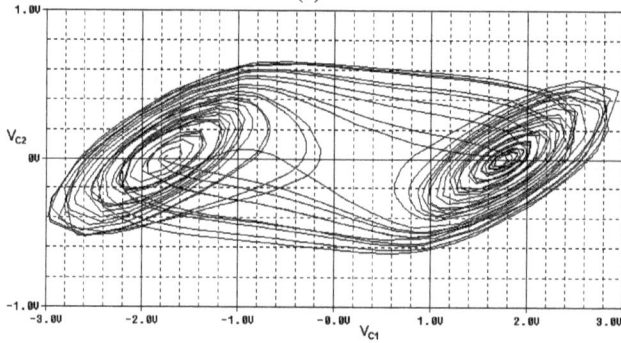

(b)

Fig. 1.21 (a) Simulations of chaotic circuit dynamics $V_{C1}(t)$, $V_{C2}(t)$ and $i_L(t)$ of Hybrid-I realization, (b) The double-scroll attractor, projection in the (V_{C2}–V_{C1}) plane.

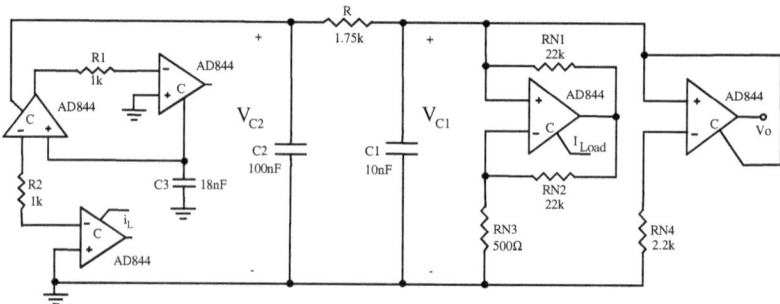

Fig. 1.22 Hybrid-II realization of Chua's circuit.

In this realization, we determined the parameter values of nonlinear resistor as $R_{N1} = R_{N2} = 22$ kΩ, $R_{N3} = 500$ Ω, and $R_{N4} = 2.2$ kΩ. And we used the same parameter values of synthetic inductor as in our Hybrid-I realization of Chua's circuit with $L = 18$ mH inductance value given by Eq. (1.14). The rest of the parameters are chosen as $C_1 = 10$ nF, $C_2 = 100$ nF, and $R = 1750$ Ω such that the circuit exhibits a double-scroll attractor. The circuit dynamics V_{C1}, V_{C2}, i_L and current output i_{Load} with a 5 kΩ load, and the double-scroll attractor obtained from PSPICE simulation experiments are shown in Fig. 1.23(a) and (b), respectively.

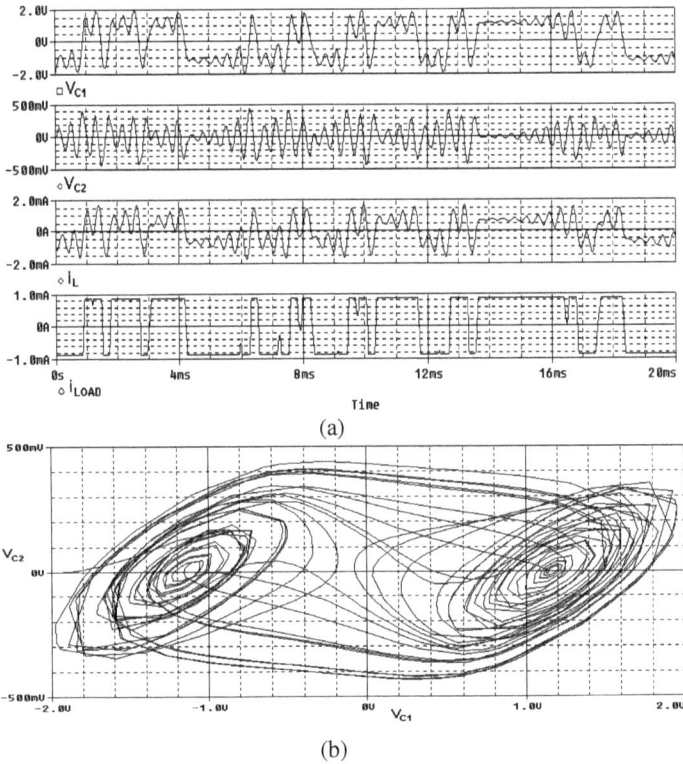

(a)

(b)

Fig. 1.23   (a) Simulations of chaotic circuit dynamics $V_{C1}(t)$, $V_{C2}(t)$, $i_L(t)$ and $i_{Load}(t)$ current output with a 5 k load of Hybrid-II realization, (b) The double-scroll attractor, projection in the (V_{C2}–V_{C1}) plane.

This realization uses only AD844-type CFOAs as the active elements. Due to the use of CFOAs for synthetic inductor and nonlinear resistor, all the state variables V_{C1}, V_{C2} and i_L are made available in a direct manner. In addition, a buffered and isolated voltage output is available, and the operating frequency can be extended to higher frequencies by a rescaling process. Moreover, a chaotic current output i_{Load} may be useful in some applications.

1.4.3 *Hybrid-III realization of Chua's circuit*

Fig. 1.24 shows a Hybrid-III realization of Chua's circuit using two AD712-type op amps biased with ±9 V, six resistors to implement a nonlinear resistor, two CMOS-based FTFN blocks, four resistors and a capacitor to implement an inductance simulator with L = 18 mH. In simulation experiments, we used the CMOS realization given in Fig. 1.14(b) for the FTFN blocks. While we determined the parameter values of the FTFN-based inductance simulator as $R_1 = R_2 = R_3 = R_4 = 1$ kΩ, and $C_3 = 18$ nF, giving an inductance of L = 18 mH according to Eq. (1.15), the rest of the parameters are chosen as $C_1 = 10$ nF, $C_2 = 100$ nF, and R = 1625 Ω such that the circuit exhibits a double-scroll attractor behavior.

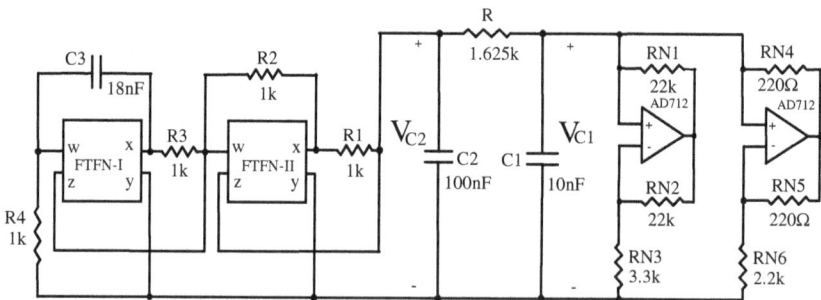

Fig. 1.24 Hybrid-III realization of Chua's circuit.

The PSPICE simulation results are shown in Fig. 1.25. Due to use of a CMOS-based FTFN topology, this realization is suitable for integrated circuit implementation. The FTFN-based inductance simulator used in this realization can also be used for simulating a floating inductance in other chaotic circuits.

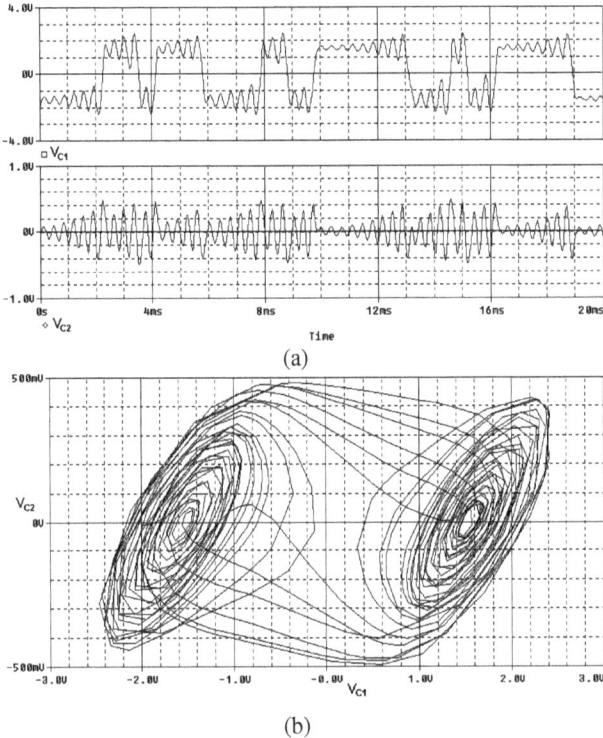

(a)

(b)

Fig. 1.25 (a) Simulations of chaotic circuit dynamics $V_{C1}(t)$ and $V_{C2}(t)$ of Hybrid-III realization, (b) The double-scroll attractor, projection in the (V_{C2}–V_{C1}) plane, in Hybrid-III realization.

1.4.4 Hybrid-IV realization of Chua's circuit

Fig. 1.26 shows the Hybrid-IV realization of Chua's circuit using two AD844-type CFOAs biased with ±9 V, four resistors to nonlinear resistor, two FTFN blocks, four resistors, and a capacitor to implement an inductance simulator with L = 18 mH. In this realization, the nonlinear resistor structure of Hybrid-III is replaced with a CFOA-based nonlinear resistor circuit. This reduces the component count, and the chaotic spectrum can be extended to higher frequencies. In addition, a buffered and isolated voltage output is made available.

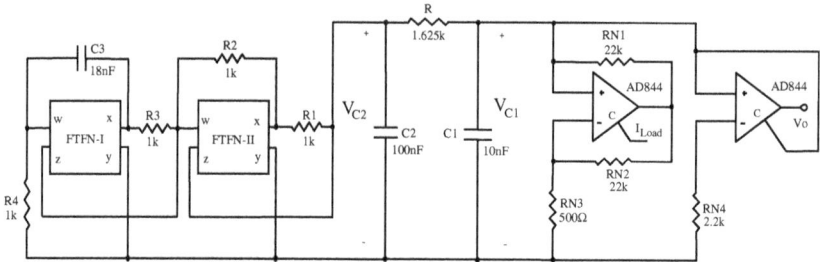

Fig. 1.26 Hybrid-IV realization of Chua's circuit.

The circuit's variables V_{C1}, V_{C2}, i_{Load} (current output with a 5 kΩ load) and a double-scroll attractor obtained by PSPICE simulations are shown in Figs. 1.27(a) and (b), respectively.

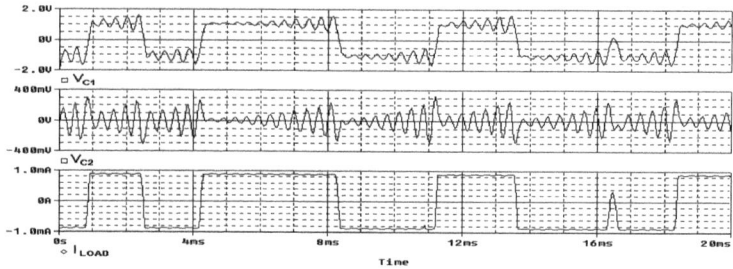

(a)

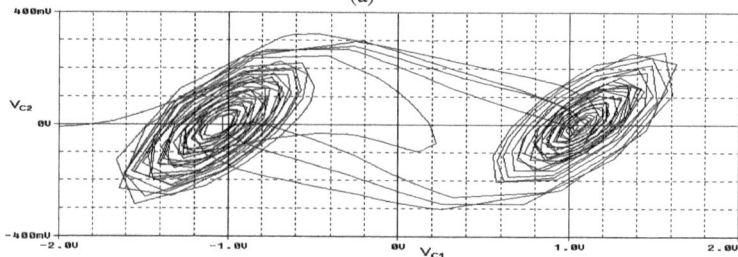

(b)

Fig. 1.27   (a) Simulations of chaotic circuit dynamics $V_{C1}(t)$, $V_{C2}(t)$ and $i_{Load}(t)$ current output with a 5 k load of Hybrid-IV realization, (b) The double-scroll attractor, projection in the $(V_{C2}-V_{C1})$ plane.

1.4.5 *Hybrid-V realization of Chua's circuit*

Fig. 1.28 shows a Hybrid-V realization of Chua's circuit using two AD844-type CFOAs biased with ±9 V, four resistors to implement nonlinear resistor, two AD712-type op amps biased with ±9 V, four resistors and a capacitor to implement a VOA-based inductance simulator with L = 18 mH. The same circuit structure in Hybrid-IV realization for nonlinear resistor was used in this realization, and the VOA-based inductance simulator's parameters are determined as $R_1 = R_2 = R_3 = R_4 = 1$ kΩ, and $C_3 = 18$ nF, giving L = 18 mH according to Eq. (1.13). The rest of the parameters are chosen as $C_1 = 10$ nF, $C_2 = 100$ nF, and R = 1650 Ω such that the circuit exhibits a double-scroll attractor. The nonidealities and parasitic effects of the VOA should be considered for obtaining the real value of the inductor. The simulation results of this realization are shown in Fig. 1.29.

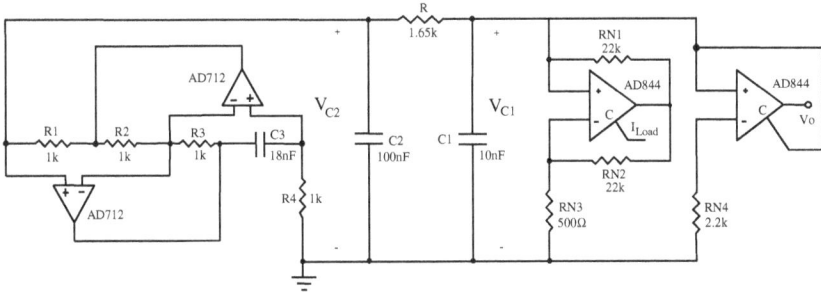

Fig. 1.28 Hybrid-V realization of Chua's circuit.

1.4.6 *Hybrid-VI realization of Chua's circuit*

Fig. 1.30 shows a Hybrid-VI realization of Chua's circuit using two AD844-type CFOAs biased with ±9 V, four resistors to implement nonlinear resistor, an AD712 VOA biased with ±9 V, four resistors and two capacitors to implement a Wien bridge–based RC configuration for the LC resonator in Chua's circuit. The parameters of the RC configuration are determined as $R_1 = 100$ Ω, $R_2 = 100$ Ω, $R_3 = 200$ Ω, $R_4 = 100$ Ω, and $C_2 = C_3 = 220$ nF. The rest of the circuit parameters are chosen as $C_1 = 2.2$ nF, and R = 1525 Ω such that the circuit exhibits a

double-scroll attractor behavior. The simulation results for this
realization are shown in Fig. 1.31.

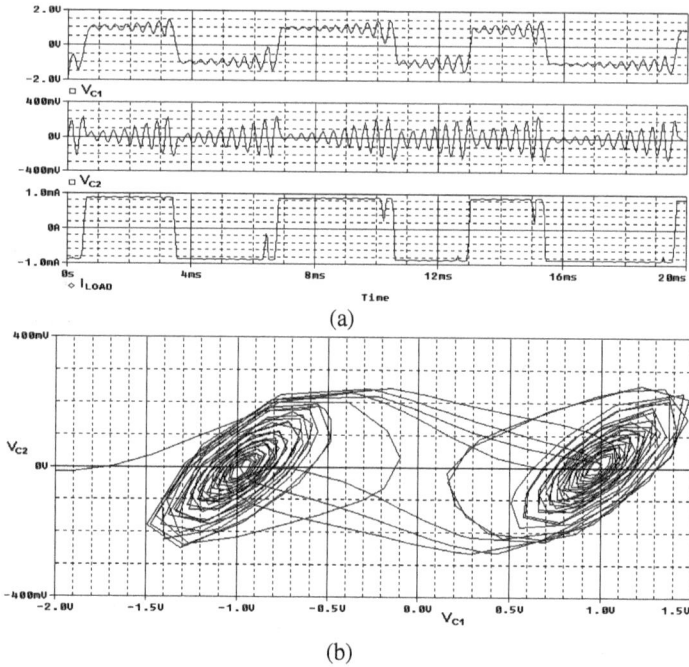

(a)

(b)

Fig. 1.29 (a) Simulations of chaotic circuit dynamics $V_{C1}(t)$, $V_{C2}(t)$ and $i_{Load}(t)$ current
output with a 5k load of Hybrid-V realization, (b) The double-scroll attractor , projection
in the (V_{C2}-V_{C1}) plane.

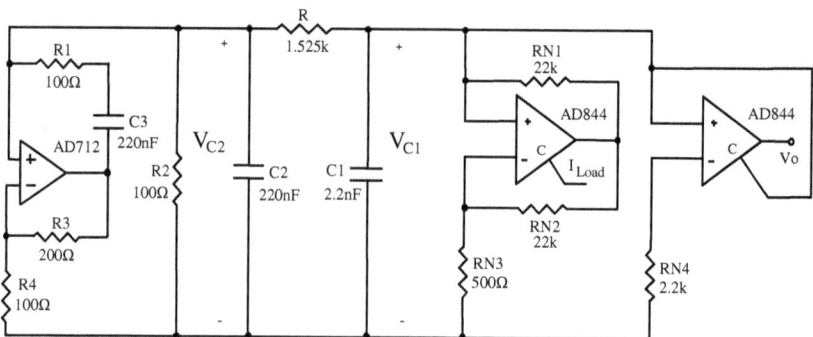

Fig. 1.30 Hybrid-VI realization of Chua's circuit.

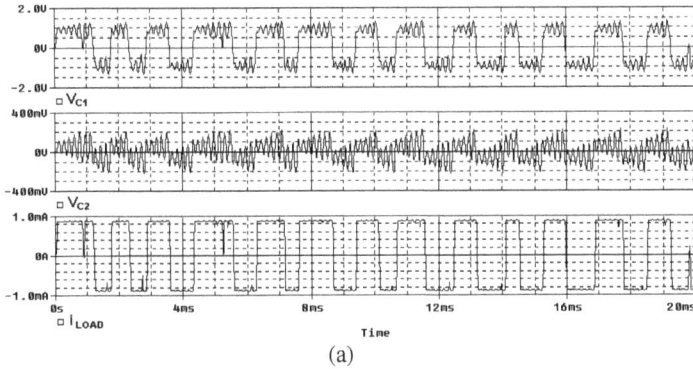

(a)

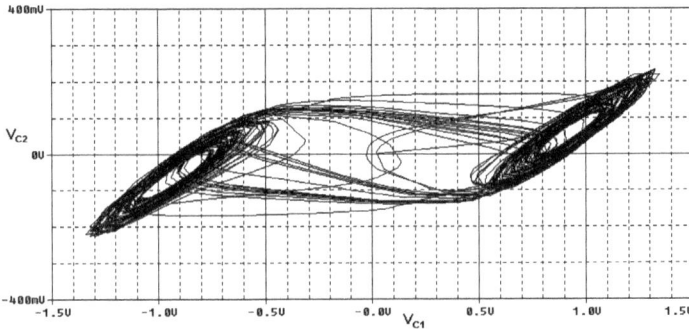

(b)

Fig. 1.31   (a) Simulations of chaotic circuit dynamics $V_{C1}(t)$, $V_{C2}(t)$ and $i_{Load}(t)$ current output with a 5 k load of Hybrid-VI realization, (b) The double-scroll attractor, projection in the (V_{C2}–V_{C1}) plane.

1.4.7 Hybrid-VII realization of Chua's circuit

Fig. 1.32 shows a Hybrid-VII realization of Chua's circuit using an FTFN-based nonlinear resistor and FTFN-based inductance simulator. In simulation experiments, CMOS realization in Fig. 1.14 was used for FTFN blocks. We determined the parameter values of the FTFN-based inductance simulator as $R_1 = R_2 = R_3 = R_4 = 1$ kΩ, and $C_3 = 18$ nF, giving an inductance of L = 18 mH according to Eq. (1.15). Other circuit elements in Fig. 1.32 are chosen as $C_1 = 10$ nF, $C_2 = 100$ nF, and R = 1650 Ω such that the circuit exhibits a double-scroll attractor behavior.

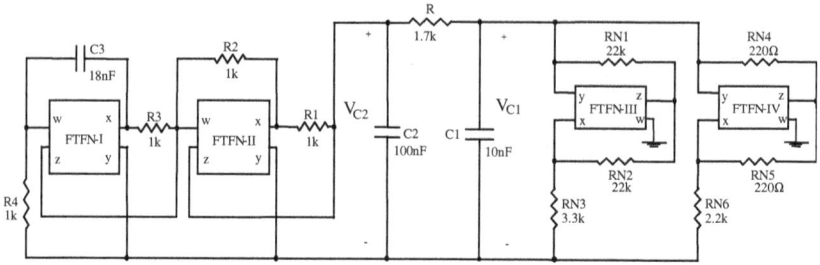

Fig. 1.32 Hybrid-VII realization of Chua's circuit.

While the chaotic waveforms and chaotic spectrum of the proposed Hybrid-VII realization of Chua's circuit are shown in Fig. 1.33, the double-scroll attractors are illustrated in Fig. 1.34.

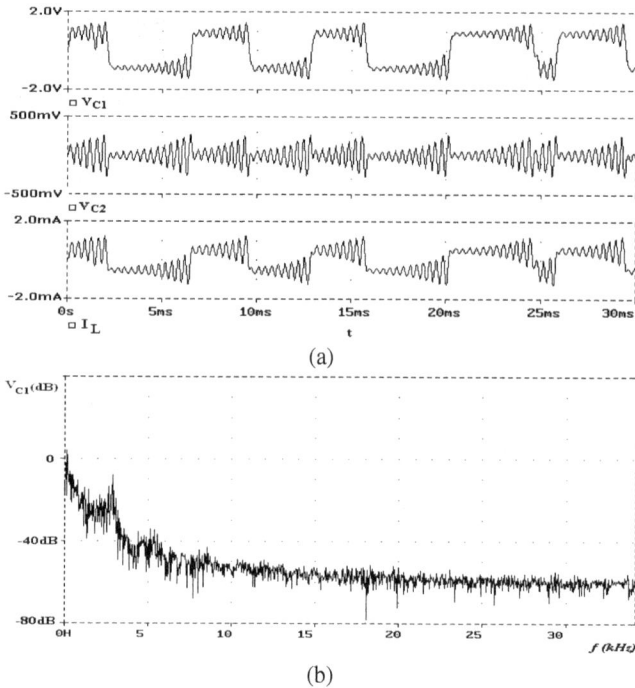

(a)

(b)

Fig. 1.33 (a) Simulations of chaotic circuit dynamics $V_{C1}(t)$, $V_{C2}(t)$ and i_L, (b) The chaotic spectrum for simulated time waveform $V_{C1}(t)$ from FTFN-based Chua's circuit.

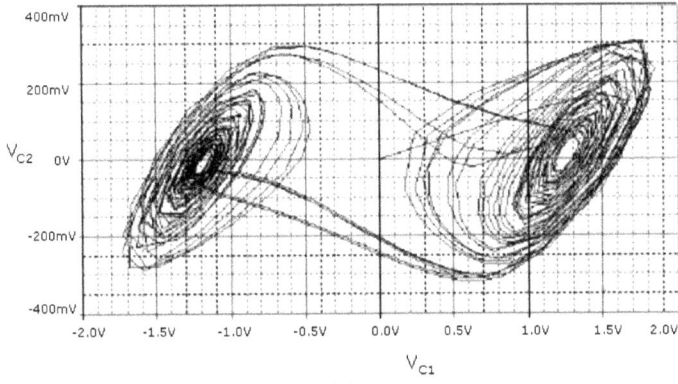

(a)

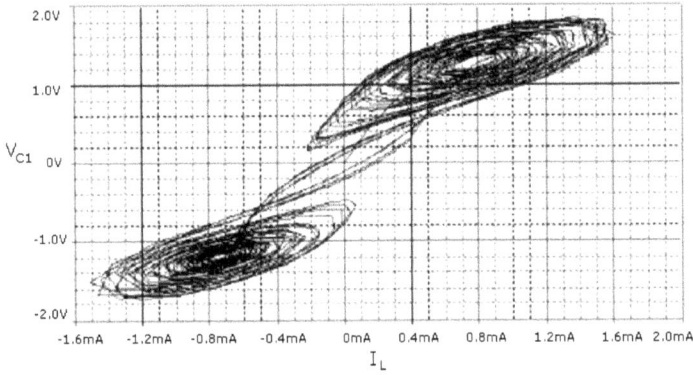

(b)

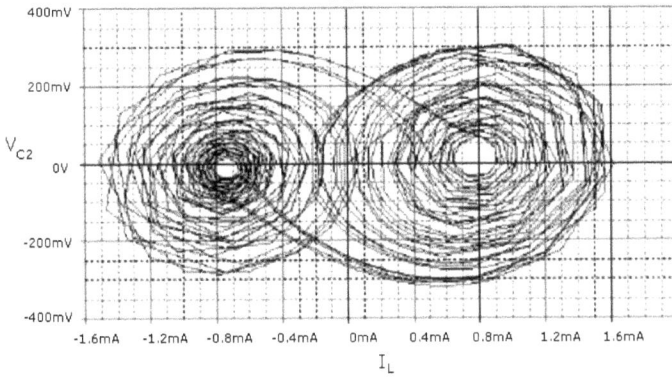

(c)

Fig. 1.34 The double-scroll chaotic attractors observed from FTFN-based Chua's circuit.

1.5 Experimental Setup of CFOA-Based Inductorless Chua's Circuit

Through PSPICE simulations, it can be seen that the Hybrid-II realization, which is constructed with a CFOA-based nonlinear resistor and CFOA-based synthetic inductor, has some advantages over the other hybrid realizations. In the Hybrid-II realization, all the state variables V_{C1}, V_{C2} and i_L are made available in a direct manner due to the use of CFOA-based topologies for both the nonlinear resistor and inductor elements. In addition, a buffered and isolated voltage output is available, while the operating frequency can be extended. Moreover, a chaotic current output i_{Load} may be useful in some applications. Hence, in this subsection we implement this Hybrid-II realization of Chua's circuit [60–61].

As a sample experimental study on Chua's circuit, the experimental implementation of a CFOA-based Chua's circuit was constructed by combining a CFOA-based Chua's diode and CFOA-based synthetic inductor. After confirming the proposed circuit's chaotic behavior by computer simulations in the previous subsections, in order to verify the circuit's operation experimentally for different frequency ranges, especially at high frequencies in which CFOA has excellent performance, we constructed a Hybrid-II realization of Chua's circuit shown in Fig. 1.22 in the form of four experimental configurations. In all experimental configurations, we fixed the parameter values of the nonlinear resistor as $R_{N1} = R_{N2} = 22$ kΩ, $R_{N3} = 2$ kΩ pot., $R_{N4} = 2.2$ kΩ, and we configured the rest of the circuit parameters as four configurations to demonstrate the circuit's behavior at different frequency ranges. This implementation uses only an AD844-type CFOA as the active element. Due to the use of CFOAs for synthetic inductor and nonlinear resistor, all the state variables V_{C1}, V_{C2} and i_L are made available in a direct manner, and a buffered voltage output is also available.

1.5.1 *Experimental results*

In the first experimental configuration, to show that the proposed circuit can exhibit the original chaotic behavior of Chua's circuit with the most studied parameters in the literature, $L = 18$ mH, $C_1 = 10$ nF, $C_2 = 100$ nF and $R = 1.7$ kΩ, we determined the circuit parameters in Fig. 1.22 as $C_1 = 10$ nF, $C_2 = 100$ nF, $R = 1.7$ kΩ, and the parameters of the CFOA-based synthetic inductor as $R_1 = R_2 = 1$ kΩ, $C_3 = 18$ nF to obtain $L = 18$ mH according to $L_{eq} = R_1R_2C_3$. In our experiments, we used a digital storage oscilloscope and spectrum analyzer interfaced to a computer for recording experimental measurements at time and frequency domains. The observed V_{C1}, V_{C2}, i_L (in the voltage form) chaotic waveforms are shown in Fig. 1.35(a)–(b). In this implementation, i_L is made available in a direct manner, and to observe this current signal at oscilloscope screen in voltage form, we used a current sensing resistor with a value of $R_{sens} = 500$ Ω. The double-scroll chaotic attractors observed in the first experimental configuration are shown in Fig. 1.36.

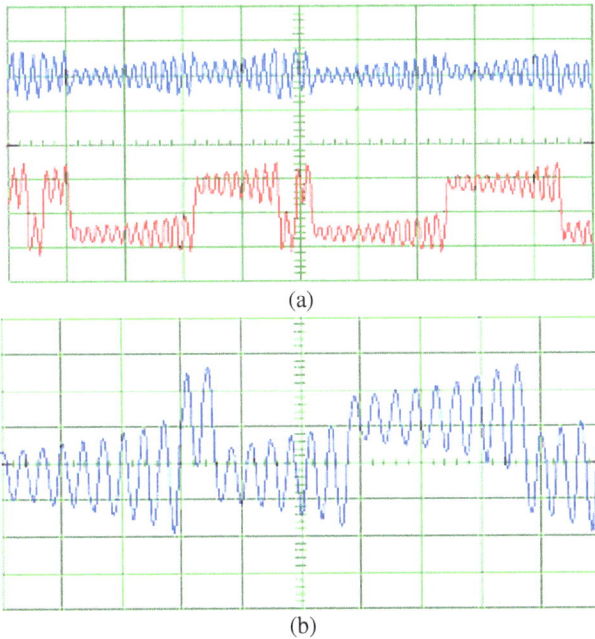

(a)

(b)

Fig. 1.35 Experimental measurements of chaotic circuit dynamics in the first experimental configuration; (a) the upper trace V_{C2} (1V/div), the lower trace V_{C1} (2V/div), time/div: 2ms/div, (b) i_L current output in the voltage form via a $R_{sens} = 500$ Ω current-sensing resistor (2V/div), time/div: 1ms/div.

(a)

(b)

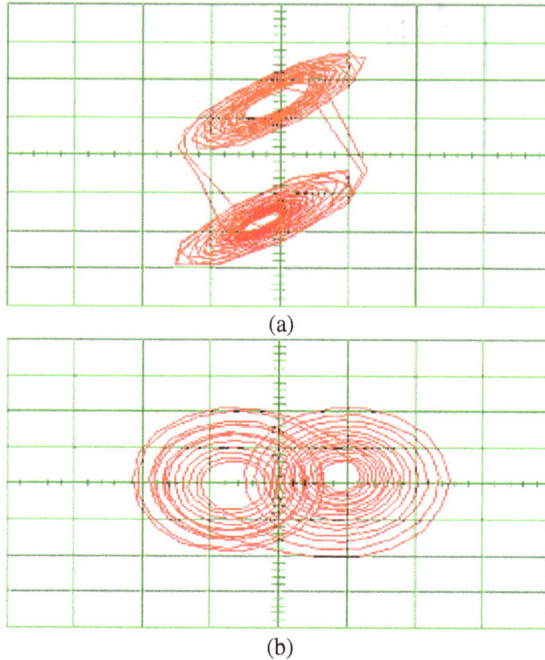

Fig. 1.36 The double-scroll attractors observed in the first experimental configuration, (a) projection in the (V_{C1}-V_{C2}) plane, x-axes: 500mV, y-axes: 1V, (b) projection in the (V_{C2}–i_L) plane, x-axes: 1V, y-axes:500mV.

These experimental results confirm that the proposed circuit is able to exhibit the original chaotic behavior of Chua's circuit with the most used parameters. After experimentally investigating the proposed circuit with the first configuration, we examined the circuit by using the second, third and fourth experimental configurations in which the synthetic inductor and capacitors are scaled down to extend the operating frequency, in order to demonstrate the high-frequency behavior of the proposed CFOA-based circuit.

In the second experimental configuration, by scaling down values of the synthetic inductor and capacitors by a factor of 50, we determined the circuit parameters in Fig. 1.22 as C_1 = 200 pF, C_2 = 2 nF, R = 1.7 kΩ, and the parameters of the CFOA-based synthetic inductor as $R_1 = R_2 = 1$ kΩ, C_3 = 0.36 nF to obtain L = 0.36 mH according to $L_{eq} = R_1R_2C_3$. For this operation mode, while the circuit dynamics V_{C1}, V_{C2}, i_L (in the voltage form) are shown in Fig. 1.37, the double-scroll attractor is illustrated in Fig. 1.38(a).

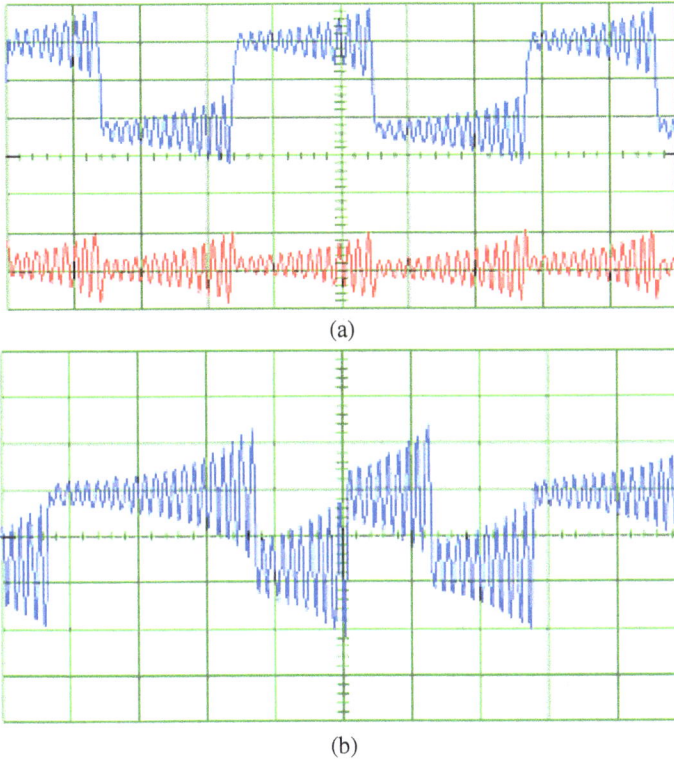

(a)

(b)

Fig. 1.37   Experimental measurements of chaotic circuit dynamics in the second experimental configuration; (a) the upper trace V_{C1} (2V/div), the lower trace V_{C2} (1V/div), time/div: 50µs/div, (b) i_L current output in the voltage form via a R_{sens} = 500Ω current-sensing resistor (1V/div), time/div: 50µs/div.

The chaotic frequency spectrum measured in the second experimental configuration is shown in Fig. 1.38(b). As shown in Fig. 1.38(b), the chaotic frequency spectrum is centered approximately around 143.75 kHz. In the third experimental configuration, by scaling down values of the synthetic inductor and capacitors by a factor of 100, we determined the circuit parameters in Fig. 1.22 as C_1 = 100 pF, C_2 = 1 nF, R = 1.7 kΩ, and the parameters of the CFOA-based synthetic inductor as R_1 = R_2 = 1 kΩ, C_3 = 0.18 nF to obtain L = 0.18 mH according to L_{eq} = $R_1 R_2 C_3$.

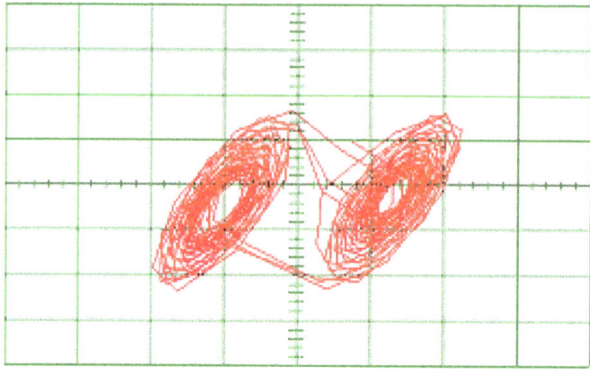

(a)

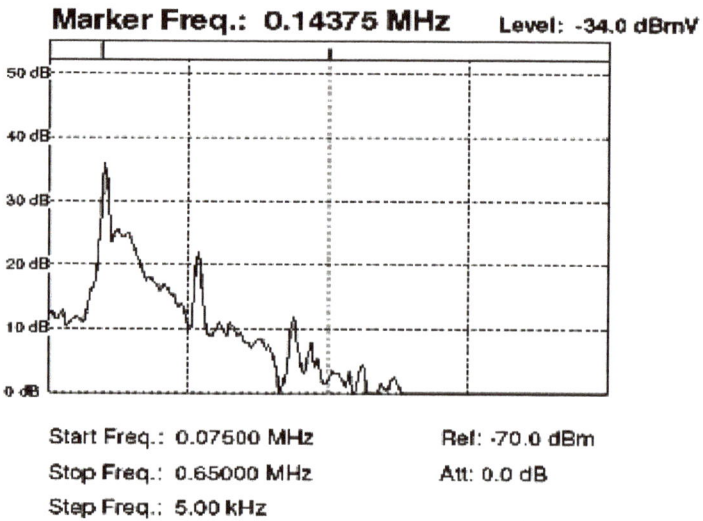

Marker Freq.: 0.14375 MHz   Level: -34.0 dBmV

Start Freq.: 0.07500 MHz   Ref: -70.0 dBm
Stop Freq.: 0.65000 MHz   Att: 0.0 dB
Step Freq.: 5.00 kHz

(b)

Fig. 1.38   For the second experimental configuration, (a) The double-scroll attractor, projection in the $(V_{C2}-V_{C1})$ plane, x-axes: 2V, y-axes: 500 mV, (b) The chaotic spectrum for measured time waveform $V_{C2}(t)$ from CFOA-based Chua's circuit.

For this high frequency oscillation mode, the circuit dynamics V_{C1}, V_{C2}, i_L (in the voltage form) are shown in Fig. 1.39, and the double-scroll attractor is illustrated in Fig. 1.40 (a). The chaotic frequency spectrum measured in the third experimental configuration is shown in Fig.

1.40(b). As shown in the figure, the operating frequency is extended and the chaotic frequency spectrum is centered approximately on 306.25 kHz.

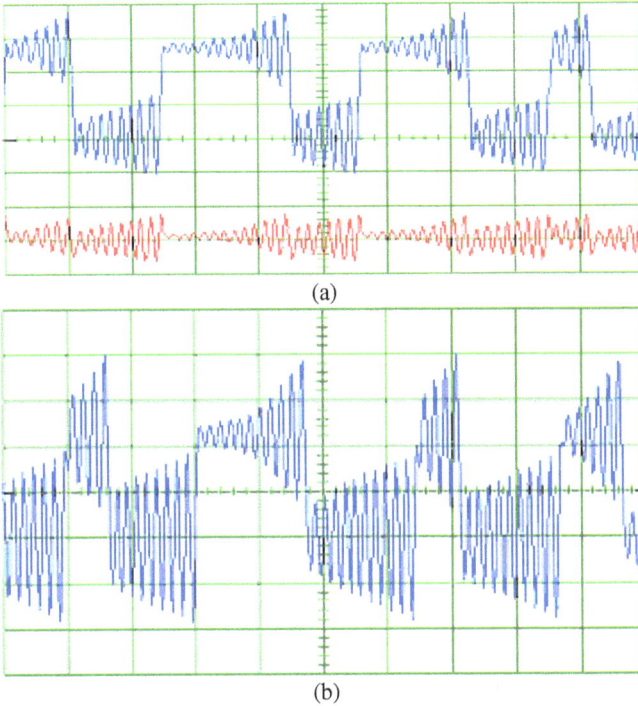

(a)

(b)

Fig. 1.39   Experimental measurements of chaotic circuit dynamics in the third experimental configuration; (a) the upper trace V_{C1} (1V/div), the lower trace V_{C2} (1V/div), time/div: 20μs/div, (b) i_L current output in the voltage form via a $R_{sens} = 500\ \Omega$ current-sensing resistor (500mV/div), time/div: 20μs/div.

In the fourth experimental configuration, by scaling down values of the synthetic inductor and capacitors by a factor of 1000, we determined the circuit parameters in Fig. 1.22 as $C_1 = 10$ pF, $C_2 = 100$ pF, R = 1.7 kΩ, and the parameters of the CFOA-based synthetic inductor as $R_1 = R_2$ = 1 kΩ, $C_3 = 0.018$ nF to obtain L = 0.018 mH according to $L_{eq} = R_1R_2C_3$. For this very high frequency operation mode, the circuit dynamics V_{C1}, V_{C2} and the double-scroll attractor are shown in Fig. 1.41, and the chaotic frequency spectrum is shown in Fig. 1.42. As shown in the figure, the operating frequency is extended to higher-frequency range and the chaotic frequency spectrum is centered approximately on 1.35 MHz.

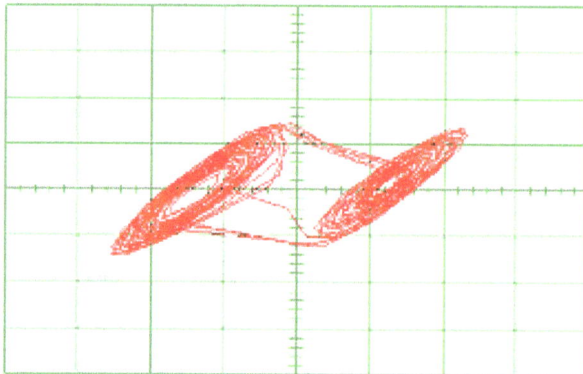

(a)

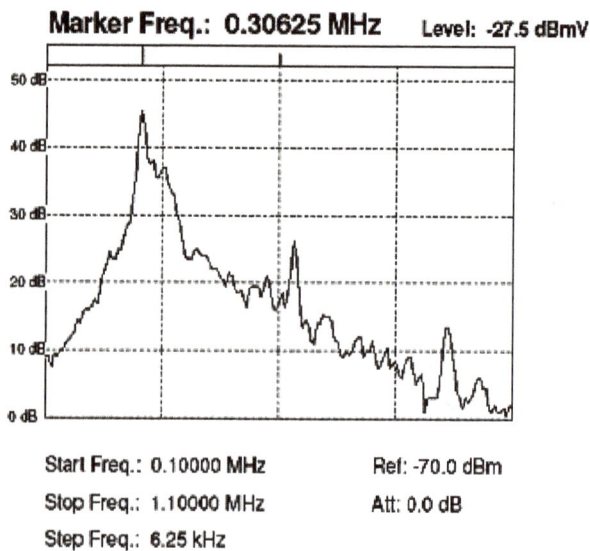

(b)

Fig. 1.40   For the third experimental configuration, (a) The double-scroll attractor, projection in the (V_{C2}–V_{C1}) plane, *x*-axes: 1 V, *y*-axes:500 mV, (b) The chaotic spectrum for measured time waveform V_{C2}(t) from CFOA-based Chua's circuit.

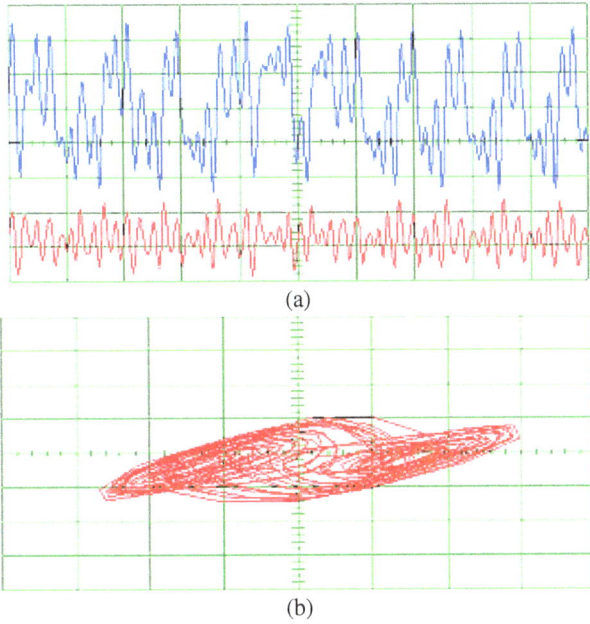

(a)

(b)

Fig. 1.41 For the fourth experimental configuration, (a) chaotic circuit dynamics; the upper trace V_{C1} (500mV/div), the lower trace V_{C2} (500mV/div), time/div: 5μs/div, (b) The double-scroll attractor, projection in the (V_{C2}–V_{C1}) plane, *x*-axes: 500 mV, *y*-axes: 500 mV.

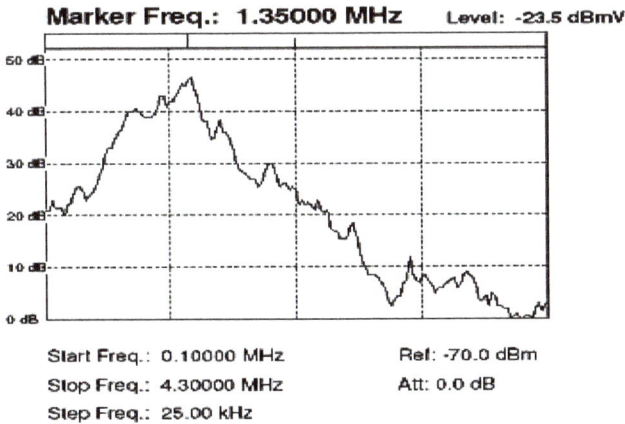

Fig. 1.42 For the fourth experimental configuration, the chaotic spectrum for measured time waveform $V_{C2}(t)$ from CFOA-based Chua's circuit.

Chapter 2

Numerical Simulation and Modeling of Chua's Circuit

Numerical simulation and mathematical modeling of a linear or nonlinear dynamic system plays a very important role in analyzing the system and predetermining design parameters prior to its physical realization. Several numerical simulation tools have been used for simulating and modeling of dynamic systems. MATLAB^{TM} [130] is one of the most powerful numerical simulation tools for solving ordinary differential equations (ODEs) which characterize dynamic systems. In addition to its numerical simulation feature, MATLAB offers a very effective graphical programming tool, SIMULINK^{TM} [130], used for modeling both linear and nonlinear dynamic systems. These MATLAB-based numerical and graphical programming solutions can also be effectively used in the process of simulation and modeling of chaotic circuits and systems that exhibit complex nonlinear dynamics [1, 11, 56, 105–106, 116, 124, 136, 142]. SIMULINK offers some advantages over the other code-based numerical programs. In the SIMULINK graphical environment, the simulation model is constructed by using basic building blocks. A set of differential equations that define the nonlinear dynamic system can thereby be modeled by interconnection of suitable function blocks that perform specific mathematical operations. This allows the user to graphically model equations and analyze the results.

This chapter presents the use of MATLAB and SIMULINK in dynamic modeling and simulation of Chua's circuit [83]. After discussing numerical simulation methods via code-based programming, we also introduce a generalized Chua's circuit model created in the SIMULINK environment.

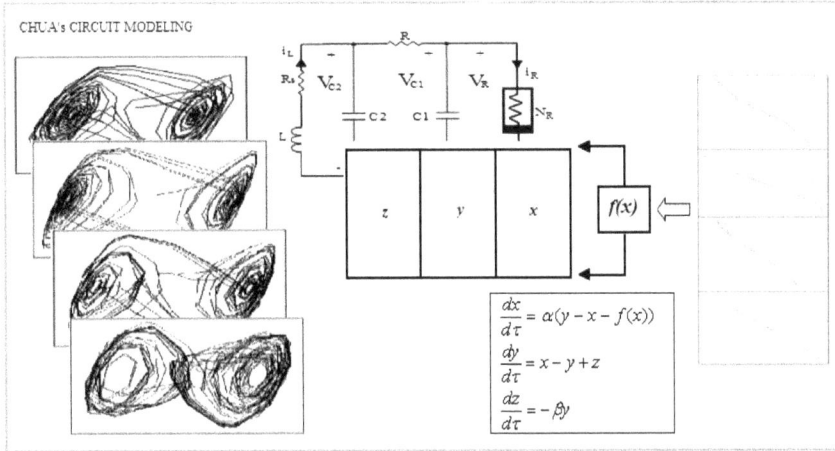

Fig. 2.1 Generalized Chua's circuit model with multiple nonlinear functions.

This generalized model consists of a main Chua's circuit's block coupled with multiple nonlinear functions via a switching mechanism. This model allows easy investigation of the roles and effects of different nonlinear functions in Chua's circuit under a unique model.

2.1 Numerical Simulations of Chua's Circuit

The original Chua's circuit is defined by the following dimensionless equations:

$$\frac{dx}{d\tau} = \alpha(y - x - f(x))$$

$$\frac{dy}{d\tau} = x - y + z \qquad (2.1)$$

$$\frac{dz}{d\tau} = -\beta y$$

Here, while x, y and z denote the state variables of the system, α and β are the system parameters. $f(x)$ represents a nonlinear function that plays an important role in the system's chaos mechanism. It has been verified that this chaotic system can be realized with many different nonlinear functions including piecewise-linear function [57], cubic function [150], piecewise-quadratic function [138] and other trigonometric functions [19]. Fig. 2.1 illustrates the generalized Chua's circuit model with multiple nonlinear functions. The most preferred functions used in Chua's circuit modeling are summarized in Table 2.1 with the typical parametric definitions.

Table 2.1 The most used nonlinear function definitions for modeling Chua's circuit.

Nonlinear Function Definition	Function Parameters
$f(x) = bx + 0.5(a-b)(|x+c| - |x-c|)$ (2.2)	$a = -1.27, b = -0.68, c = 1$
$f(x) = h_1 x - h_2 x^3$ (2.3)	$h_1 = -1.27, h_2 = -0.0157$
$f(x) = -a\tanh(bx)$ (2.4)	$a = 2, b = 0.38$
$f(x) = d_1 x + d_2 x|x|$ (2.5)	$d_1 = -8/7, d_2 = 4/63$

For numerical simulation of chaotic systems defined by a set of differential equations such as Chua's circuit, different integration techniques can be used in simulation tools. In the MATLAB numerical simulations, ODE45 solver yielding a fourth-order Runge-Kutta integration solution has been used. For introducing the numerical solutions in MATLAB for Chua's circuit defined by differential equations in Eq. (2.1), we give MATLAB codes in Table 2.2 using four different nonlinear functions whose parametric details are listed in Table 2.1. In these numerical simulation profiles, first a general function description for Chua's circuit equations in Eq. (2.1) must be defined. This MATLAB code formatted function file is shown in Table 2.2 entitled with the "chuaequation.m" m-file name. And then a MATLAB m-file including ODE45 function is executed for solving the equations defined by the function m-file. This m-file is given in Table 2.3. It is noted that in solution files some graphical plot commands have been used for plotting the solutions. The other simulation files for Chua's circuit realizations using other nonlinear functions can be prepared in a similar manner.

Table 2.2 MATLAB "chuaequation.m"—function file for Chua's circuit.

MATLAB "chuaequation.m"—Function file
function dx=chuaequation(t, x)
% This is the function file 'chuaequation.m'
% Inputs are: t=time, x=[x(1); x(2); x(3)]. These inputs represent x, y and z state
%variables in Eq.(2.1).
alpha=10; beta=14.87;
% alpha and beta are constants.
a=-1.27; b=-0.68; c=1;
% h1=-1.27; h2=-0.0157;
% k1=-8/7; k2=4/90;
% d1=-8/7; d2=4/63;
% a, b, c, h1, h2, k1, k2, d1 and d2 are changeable nonlinear function parameters.
dx = zeros (3,1); % dx is a 3x1 zero matrix.
f=b*x(1)+(1/2)*(a-b)*(abs(x(1)+c)-abs(x(1)-c)); % The first nonlinear function.
%f=h1*x(1)-(h2*x(1)*x(1)*x(1)); % The second nonlinear function.
%f=k1*x(1)+(k2*x(1)*x(1)*x(1)); % The third nonlinear function.
%f=d1*x(1)+(d2*x(1)*abs(x(1))); % The fourth nonlinear function.
dx(1) = alpha*(x(2)-x(1)-f); % dx(1) represents $\dfrac{dx}{d\tau}$.
dx(2) = x(1)-x(2)+x(3); % dx(2) represents $\dfrac{dy}{d\tau}$.
dx(3) = -beta*x(2); % dx(3) represents $\dfrac{dz}{d\tau}$.

According to these numerical simulations, the circuit's chaotic dynamics and double-scroll attractors are shown in Fig. 2.2. These numerical solutions can be compared with the former results obtained from PSPICE simulations and laboratory experiments presented in the first chapter.

Table 2.3 MATLAB executing file for Chua's circuit.

MATLAB "chuaequation_run.m"—Executing file
[t, x] = ode45('chuaequation', [0 100], [0 0 1]); % executes ode45 solver.
% t is time vector and x is solution matrix. [0 100] is time span [t_o t_{final}].
% [0 0 1] is initial conditions of x(1), x(2) and x(3) respectively.
figure; % open a new figure window
subplot (3,1,1); % divide the graphics window into three sub windows and plot in
% the first window.
plot (t, x(:,1),'–'); ylabel('x'); % plot in the first sub window and label **y**-axis as *x*.
subplot(3,1,2); % get figure sub window 2 ready.
plot(t, x(:,2),'–'); ylabel('y'); % plot in the second subwindow and label y-axis as *y*.
subplot(3,1,3); % get figure sub window 3 ready.
plot(t, x(:,3),'–'); ylabel('z'); xlabel('t'); % plot in the third sub window and label
% y-axis as *z*.
figure; % open a new figure window
plot(x(:,1), x(:,2),'–'); ylabel('y'); xlabel('x'); axis('tight')
% plot x(2) according % to x(1) and label x and y axes as x and y and set the
% length scales of the two axes to be tight.

2.2 Simulation and Modeling of Chua's Circuit in SIMULINK

In addition to MATLAB's code-based solution, there is a very effective tool used for modeling linear and nonlinear dynamic systems: MATLAB/SIMULINK. SIMULINK offers some advantages over the other code-based numerical programs. In the SIMULINK graphical environment, the simulation model is constructed by using basic function blocks. A set of differential equations that define the nonlinear dynamic system can thereby be modeled by interconnection of suitable function blocks that perform specific mathematical operations. Dynamic systems can be analyzed as continuous or discrete time systems in SIMULINK. Chaotic systems can be analyzed as continuous time systems which model several alternative building blocks such as the transfer functions, integration blocks and state-space blocks shown in Fig. 2.3.

(a)

(b)

(c)

(d)

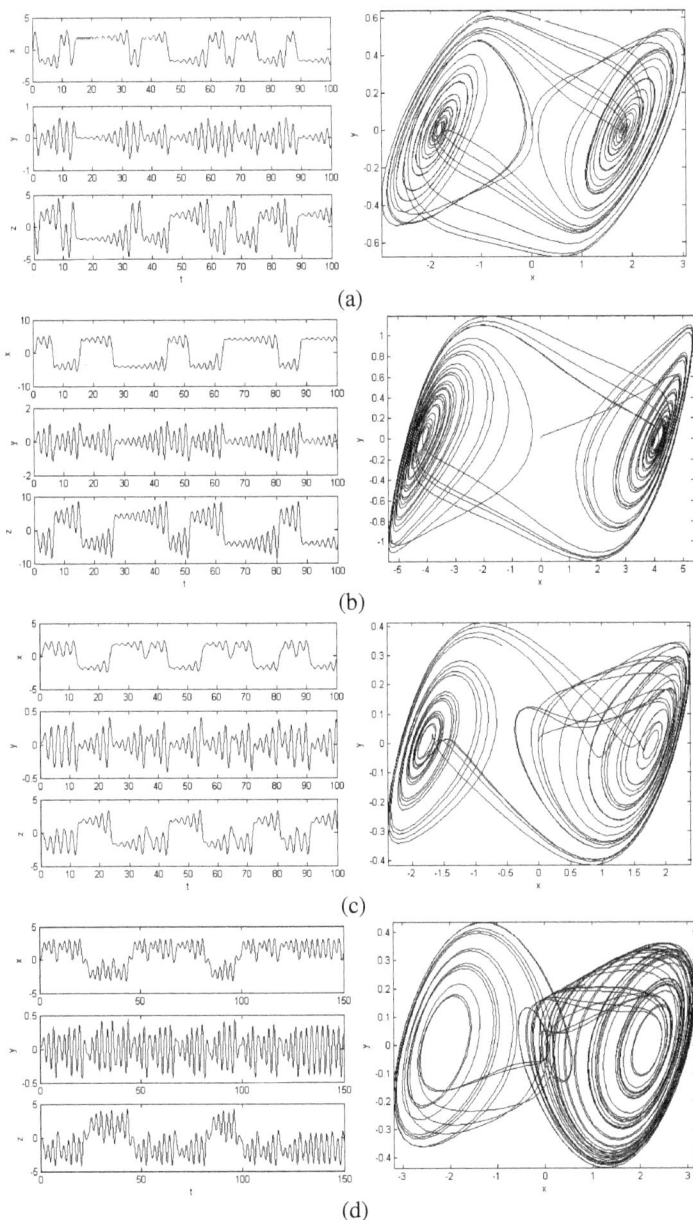

Fig. 2.2 Chua's circuit's chaotic dynamics and double-scroll attractors according to (a) $f(x) = bx + 0.5(a-b)(|x+c| - |x-c|)$, (b) $f(x) = h_1 x - h_2 x^3$, (c) $f(x) = k_1 x + k_2 x^3$, and (d) $f(x) = d_1 x + d_2 x |x|$ nonlinear functions.

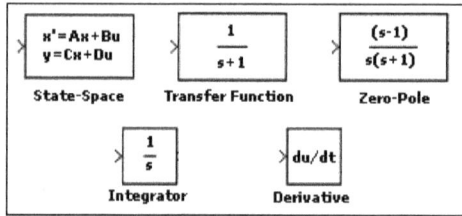

Fig. 2.3 Some building blocks representing continuous time dynamic systems.

In this section, a generalized SIMULINK model of Chua's circuit is given and the usage of SIMULINK blocks for constructing Chua's circuit with different nonlinear functions is shown. A SIMULINK model for Chua's circuit with the piecewise-linear nonlinear function is shown in Fig. 2.4.

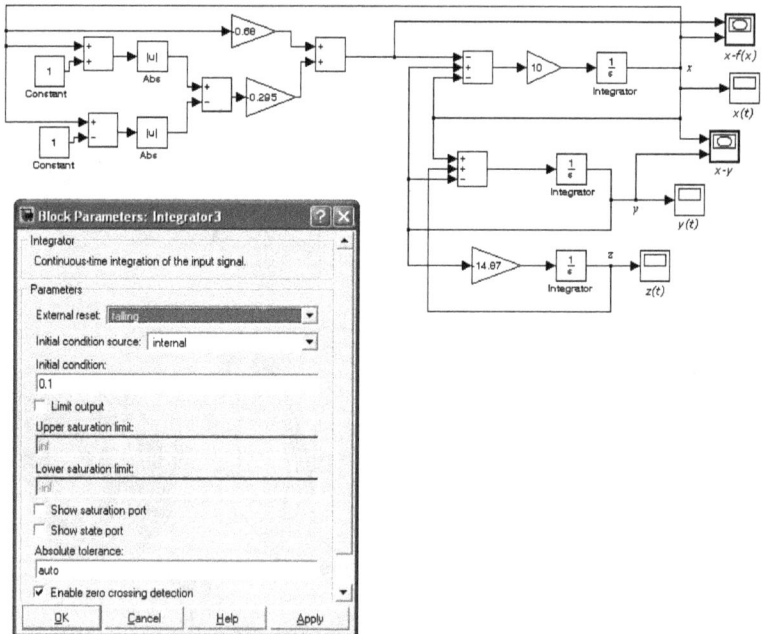

Fig. 2.4 SIMULINK model of Chua's circuit with piecewise-linear nonlinear function defined by Eq. (2.2) in Table 2.1.

As shown in the figure, system state-variables x, y and z are represented at the output of INTEGRATOR blocks. Fig. 2.4 also shows typical parameter settings of a sample integrator block via a dialog box. As stated earlier, there is an alternative construction for the main block of Chua's circuit model in SIMULINK. It can also be constructed by state-space block with appropriate parameter settings. N_{R1} defined by Eq. (2.2) in Table 2.1 is a three-segment piecewise-linear function, and it is constructed by using gain blocks, absolute blocks, adding and constant blocks. The simulation results can be monitored as time/frequency-domain response and X-Y phase portrait illustrations in SIMULINK. In addition to obtaining the simulation results including double-scroll attractors and time domain chaotic response, it is also possible to obtain the dc characteristic of each nonlinear function block. In Fig. 2.5, a sample SIMULINK modeling for determining the dc characteristic of piecewise-linear nonlinear function block (NR1) used in Fig. 2.4 is depicted. As shown in the figure for obtaining the dc characteristic of the related nonlinear function block in SIMULINK, a ramp generator function is applied to the input of the nonlinear function block.

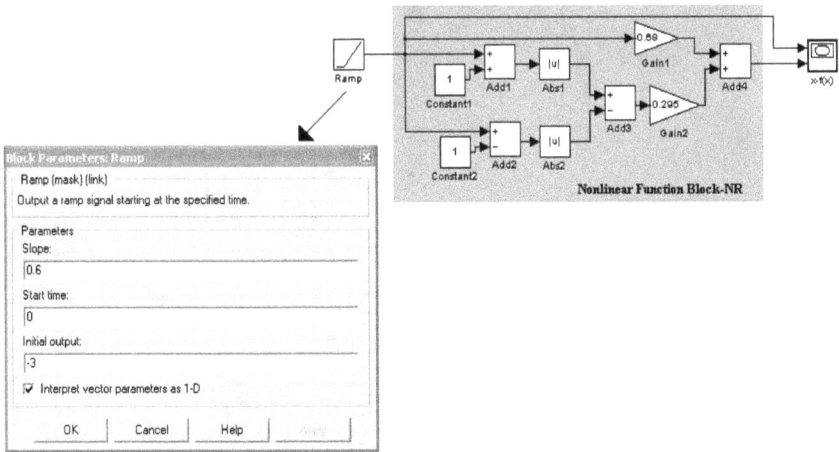

Fig. 2.5 A SIMULINK model for determining dc characteristic of nonlinear function block (NR1) in Chua's circuit model in Fig. 2.4.

Typical block parameter settings of ramp generator block are also demonstrated in Fig. 2.5. Chaotic dynamics, dc characteristic of the nonlinear function and double-scroll chaotic attractor illustration obtained from Chua's circuit model in Fig. 2.4 are shown in Fig. 2.6. Similarly, Chua's circuit model defined by Eq. (2.1) can be realized by using other nonlinear functions defined in Table 2.1. SIMULINK models of these alternative realizations are depicted in Fig.2.7.

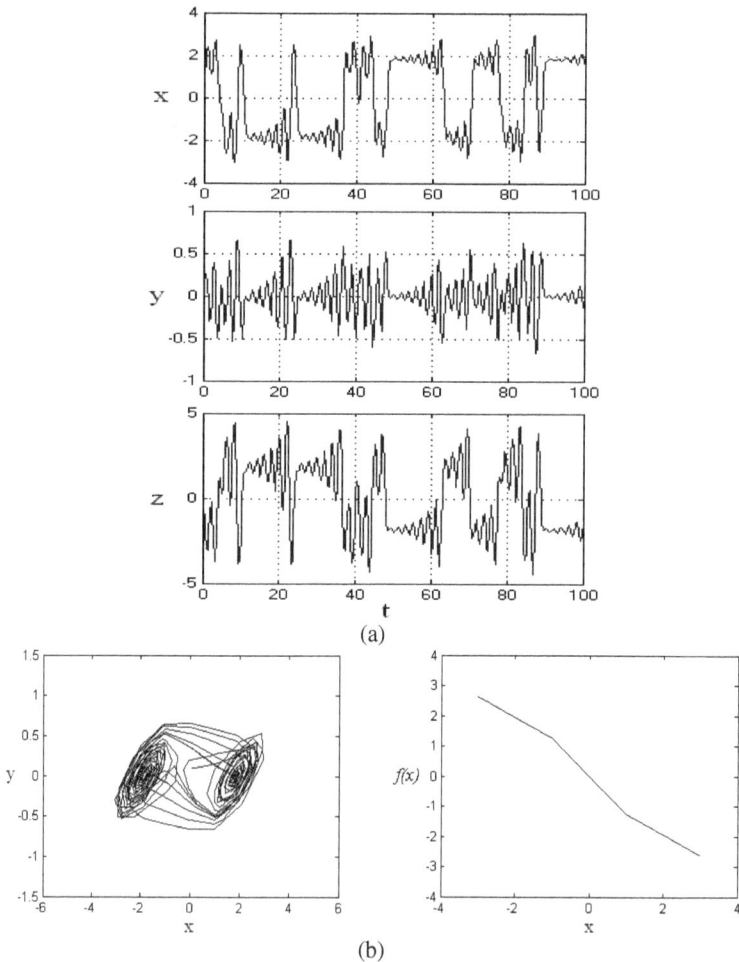

(a)

(b)

Fig. 2.6 SIMULINK simulation results of Chua's circuit model in Fig. 2.4 showing chaotic dynamics, chaotic attractor and dc characteristic of nonlinear function block.

(a)

(b)

(c)

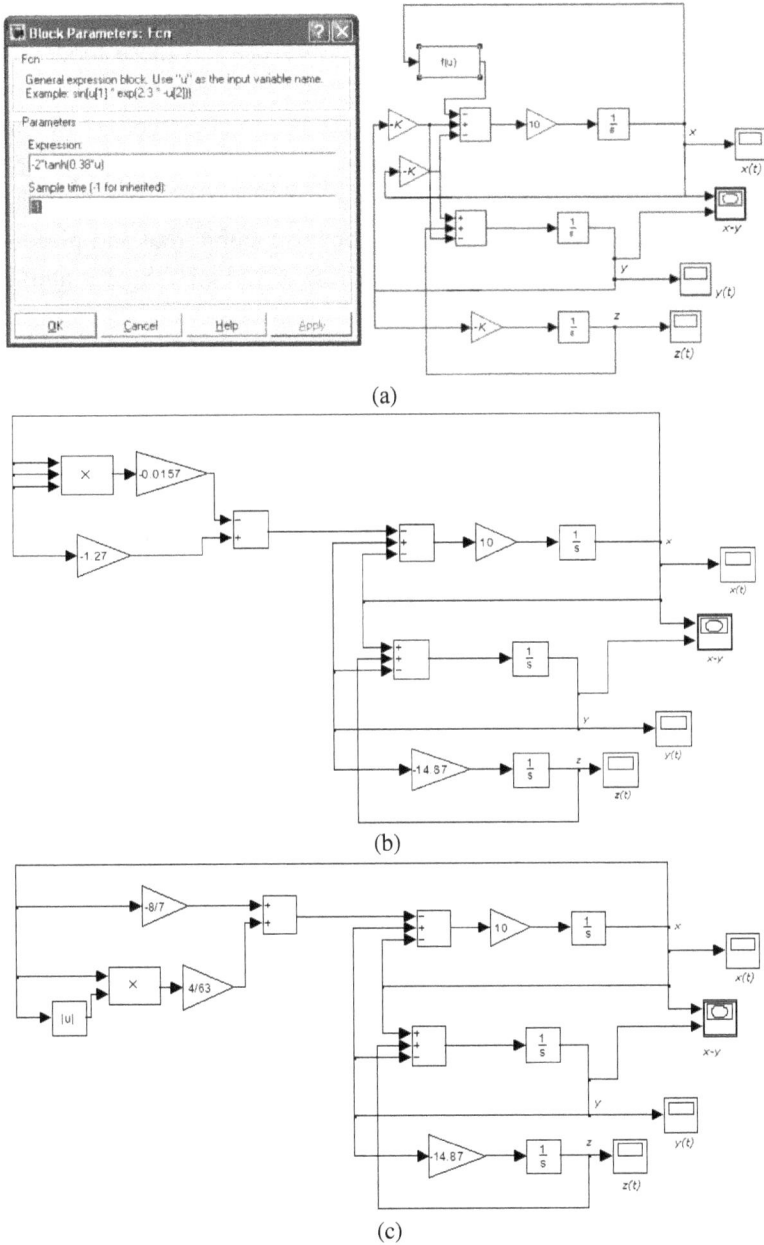

Fig. 2.7 SIMULINK models of Chua's circuit using (a) trigonometric nonlinear function, (b) cubic-like nonlinear function and (c) piecewise-quadratic nonlinear function.

As shown in models, the nonlinear functions including trigonometric, cubic-like and piecewise-quadratic functions used in these models have been constructed by using different SIMULINK blocks. For example, trigonometric nonlinear function has been constructed by using FUNCTION block. Fig. 2.7(a) shows the parameter settings of this block. SIMULINK simulation results obtained from alternative Chua's circuit models in Fig. 2.7 using trigonometric, cubic-like and piecewise-quadratic nonlinear functions are illustrated in Fig. 2.8, 2.9 and 2.10, respectively.

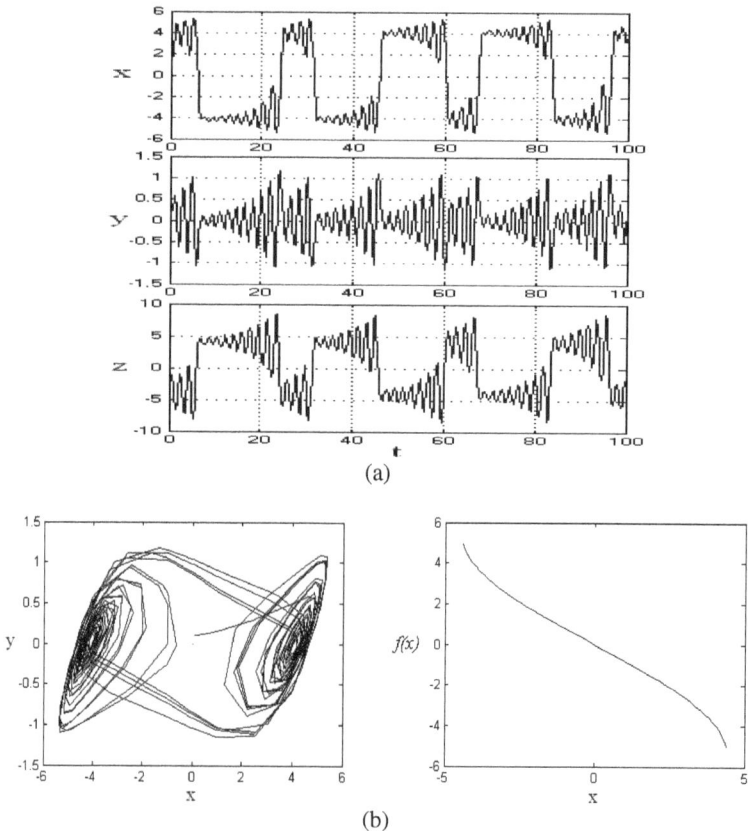

(a)

(b)

Fig. 2.8   SIMULINK simulation results of Chua's circuit model in Fig. 2.7(a) showing chaotic dynamics, chaotic attractor and dc characteristic of nonlinear function block.

(a)

(b)

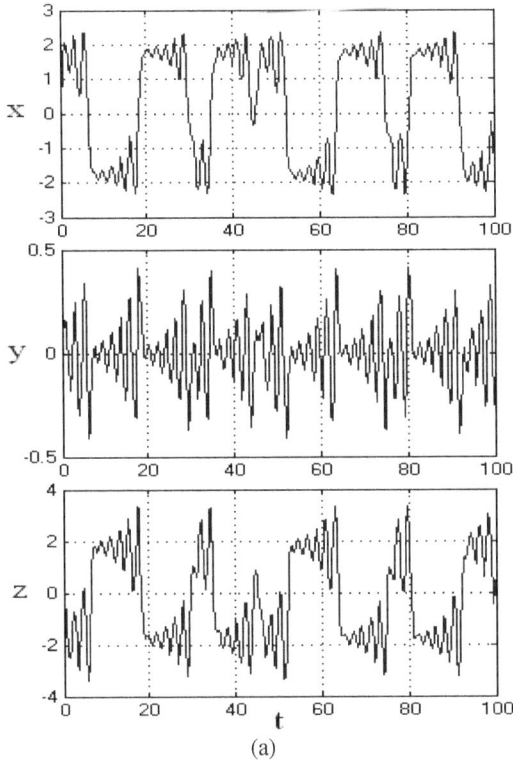

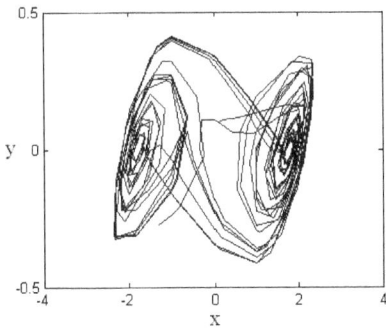

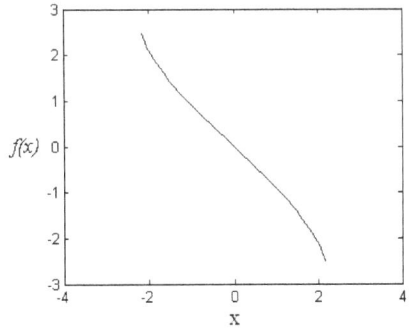

Fig. 2.9 SIMULINK simulation results of Chua's circuit model in Fig. 2.7(b) showing chaotic dynamics, chaotic attractor and dc characteristic of nonlinear function block.

(a)

(b)

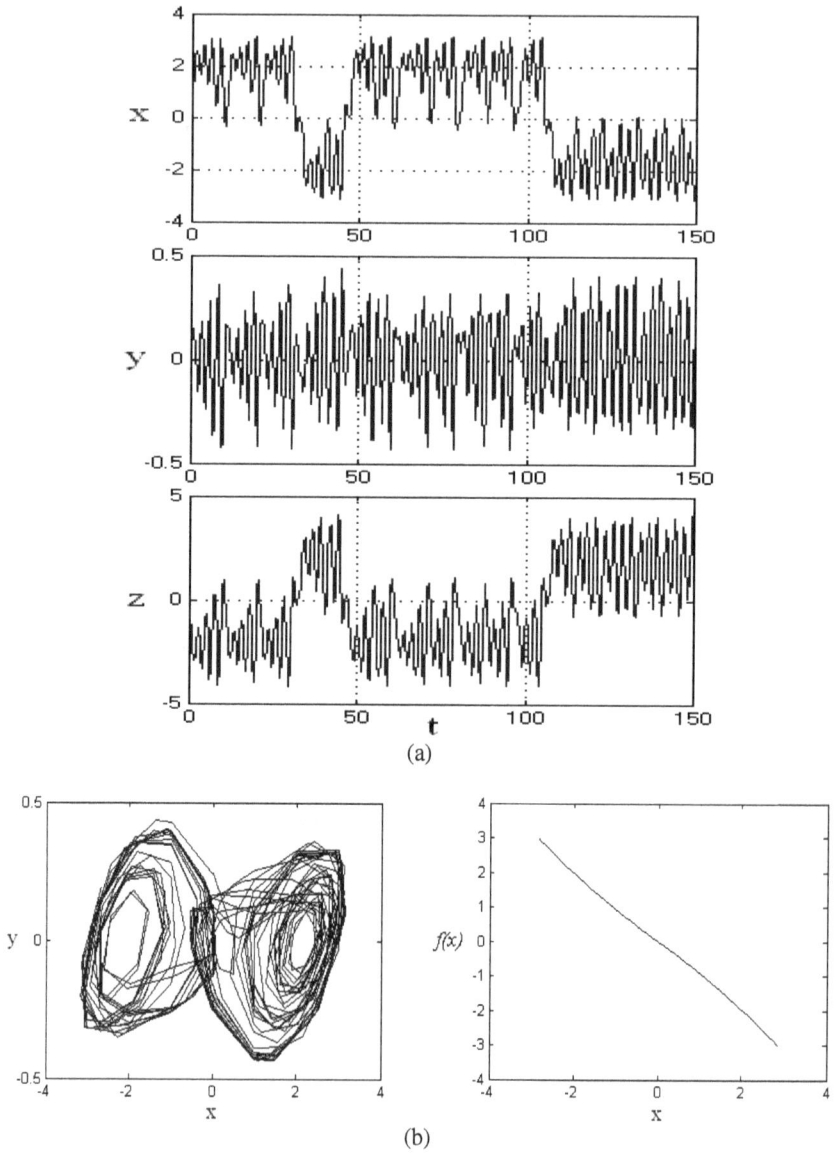

Fig. 2.10 SIMULINK simulation results of Chua's circuit model in Fig. 2.7(c) showing chaotic dynamics, chaotic attractor and dc characteristic of nonlinear function block.

It is noted that except for Chua's circuit model in Fig. 2.7(a) using trigonometric nonlinear function, other Chua's circuit models have the same main system part including integrator blocks in which system state variables are represented. Therefore, these models using the same system part can be investigated in a generalized SIMULINK model. This generalized Chua's circuit model defined by Eq. (2.1) and the nonlinear function definitions in Table 2.1 is shown in Fig. 2.11. Piecewise-linear function can also be constructed using activation function block as shown in the SIMULINK model in Fig. 2.11.

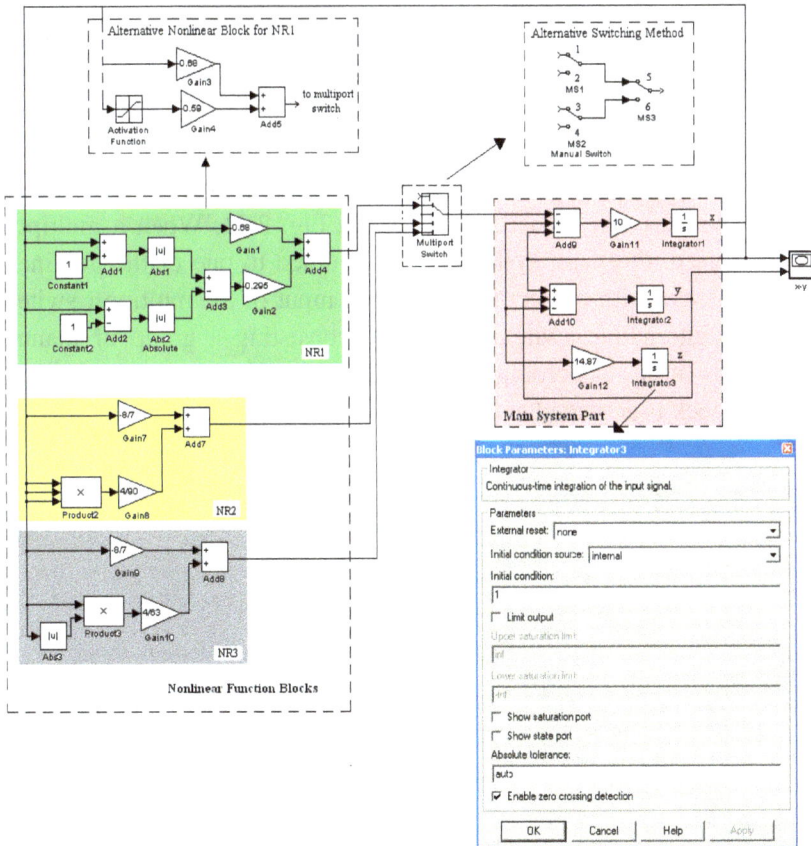

Fig. 2.11 A generalized SIMULINK model for Chua's circuit with multiple nonlinear functions.

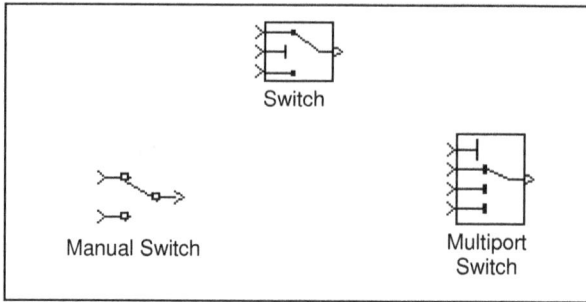

Fig. 2.12 Some nonlinear operator blocks representing switches.

Different type switches are included in SIMULINK as nonlinear operators. Some of them are illustrated in Fig. 2.12. Manual switches are useful when trying simulations with different cases. For coupling between the main system part and nonlinear function blocks of the generalized Chua's circuit model, a single multiport switch or a group of manual switches can be used as shown in Fig. 2.11. When a multiport switch is used, the main system part is coupled to one of the nonlinear function blocks with respect to the control input of the multiport switch. When the alternative switching method consisting a group of manual switches is used, according to the switch positions each nonlinear function block is activated and coupled with the main block.

Chapter 3

Programmable and Reconfigurable Implementations of Chua's Circuit Model

Among the nonlinear dynamic systems producing chaos, Chua's circuit model has very high potential for programmable and reconfigurable design. As mentioned in the earlier chapters, the advantages of Chua's circuit are the ability to tune its parameters to create a rich variety of bifurcation and chaos and a modular design based on a fixed main system block and many different changeable nonlinear function blocks including a piecewise-linear function, a cubic function, a piecewise-quadratic function and other trigonometric functions. A significant number of discrete circuit hardware components is needed to implement and test the nonlinear functions that compose Chua's circuit. Programmable devices are becoming increasingly popular as an alternative to these discrete implementations. Both digital circuits and analog circuits are now available in programmable form. Analog systems can be flexibly changed through software, making it easier and cheaper to implement new designs. A field programmable analog array (FPAA) is a programmable device for analog circuit design that can be effectively used to reconfigure implementations of Chua's circuit. This device is more efficient, simpler, and more economical than a combination of discrete components (*e.g.*, op amps, comparators, analog multipliers) that otherwise would be used to model Chua's circuit. The FPAA design approach can be used to obtain a fully programmable Chua's circuit, with the ability to change circuit parameters on the fly, to model a chaotic system using nonlinear function blocks, and to allow rapid model changes.

In this chapter, we introduce FPAA-based Chua's circuit models that use different nonlinear functions in a programmable and reconfigurable form. First, we introduce the essentials of the FPAA device, and then we discuss FPAA-based implementations of Chua's circuit.

3.1 FPAA: General Concepts and Design Approach

A field programmable analog array is a programmable device for implementing a rich variety of systems, including analog functions. It can be dynamically reconfigured by changing component values and interconnections. Thus a design modification or a new design can be downloaded to a FPAA without the resetting the system. In addition, FPAAs provide more efficient and economical solutions for dynamic analog system designs, allowing analog chaos generators to be implemented at a lower cost, in a much smaller size and with increased reliability and component stability [5, 6, 8, 12–13, 15, 32, 51, 81]. FPAA technology enables flexible and rapid prototyping of analog circuits by programming a matrix of internal elements called configurable analog blocks (CABs). A FPAA device typically includes two or more CABs and a configuration logic unit. The CAB consists of op-amps, an array of switches, and a capacitor bank, and the configuration logic unit consists of clock sources, a shift register, and memory. A typical FPAA block diagram with a CAB structure [5, 36] is depicted in Fig. 3.1.

In FPAA structures, switched-capacitor technology is used to implement various analog functions in a CAB. For example, switched capacitors can emulate a resistor and can easily change resistor values with reprogramming. Other circuit configurations can be obtained by adjusting the switches between the CAB's components.

FPAA devices are accompanied by a software development tool with a serial interface that allows analog circuit designs to be developed and tested in a computer environment prior to their download to the FPAA hardware. The software includes predefined function blocks called CAMs. The required analog functions can be constructed by dragging and dropping CAMs on a drawing board. There are several commercial FPAA products produced by different IC manufacturers. Our experimental studies use the AN221E04 type FPAA produced by Anadigm, Inc. [5], which is integrated on the AN221K04 development board.

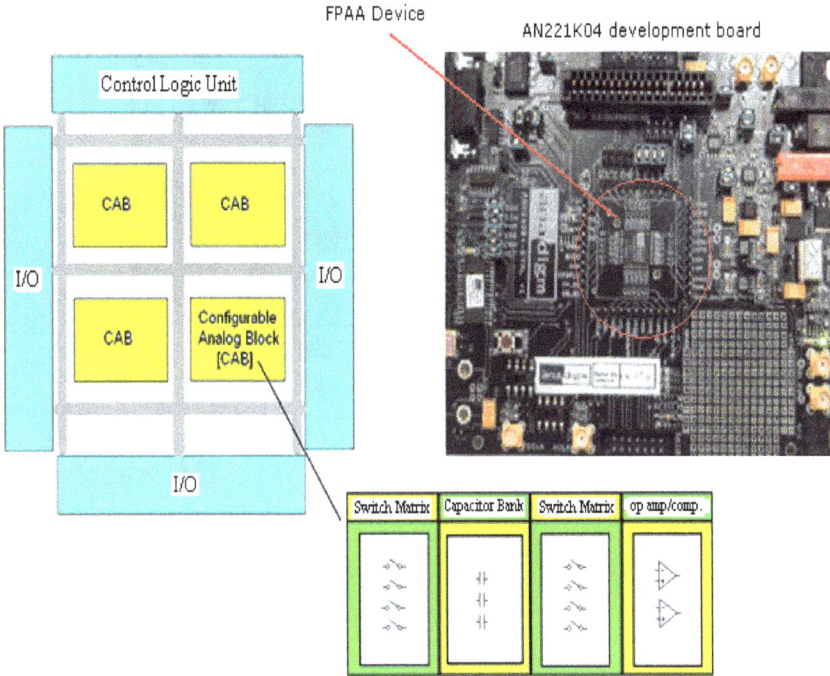

Fig. 3.1 FPAA development board and a typical FPAA block diagram.

The Anadigm Designer2 software tool, illustrated in Fig. 3.2, controls the configuration of the AN221K04 board. A partial list of the CAMs available in the Anadigm Designer2 software is provided in Table 3.1, showing that nearly all of the basic analog building blocks required for analog circuit design are available. These building blocks are well documented in the software tool. Because SUMFILTER blocks are extensively used in the circuit implementations presented in this chapter, we use their documentation details as an example of the available documentation support.

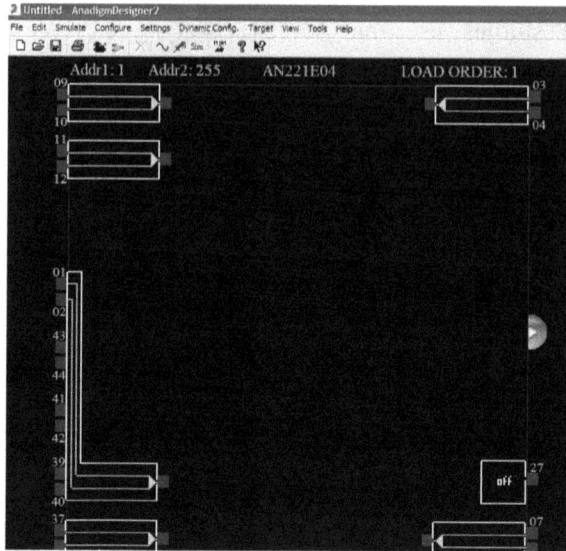

Fig. 3.2 The general appearance of the Anadigm Designer2 software.

Table 3.1 CAMs list avaliable in the Anadigm Designer2 software.

CAM	Description
ADC-SAR	Analog to Digital Converter (SAR)
Comparator	Comparator
Differentiator	Inverting Differentiator
Divider	Divider
FilterBilinear	Bilinear Filter
FilterBiquad	Biquadratic Filter
FilterBiquadPol...	Biquadratic Filter with Independent Pole/Zero
FilterDCBlockLP	DC Blocking HPF with Optional LPF
FilterLowFreqBi...	Low Corner Frequency Bilinear LPF (External...
FilterVoltageCo...	Voltage Controlled Filter
GainHalf	Half Cycle Gain Stage
GainHold	Half Cycle Inverting Gain Stage with Hold
GainInv	Inverting Gain Stage
GainLimiter	Gain Stage with Output Voltage Limiting
GainPolarity	Gain Stage with Polarity Control
GainSwitch	Gain Stage with Switchable Inputs
GainVoltageCo...	Voltage Controlled Variable Gain Stage
Hold	Sample and Hold
HoldVoltageCo...	Voltage Controlled Sample and Hold
Integrator	Integrator
Multiplier	Multiplier
OscillatorSine	Sinewave Oscillator
PeakDetect2	Peak Detector
PeakDetectExt	Peak Detector (External Caps)
PeriodicWave	Arbitrary Periodic Waveform Generator
RectifierFilter	Rectifier with Low Pass Filter
RectifierHalf	Half Cycle Rectifier
RectifierHold	Half Cycle Inverting Rectifier with Hold
SquareRoot	Square Root
SumBiquad	Sum/Difference Stage with Biquadratic Filter
SumDiff	Half Cycle Sum/Difference Stage

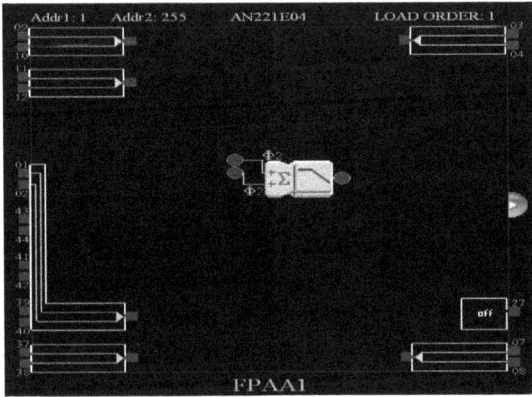

(a)

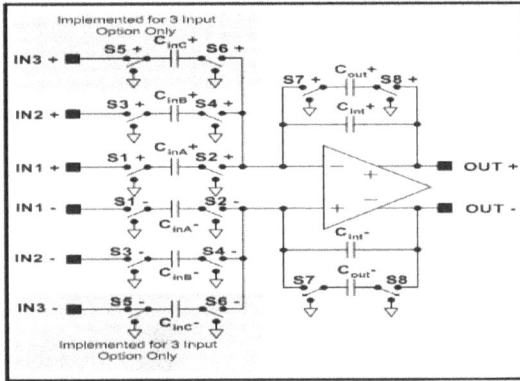

(b)

Fig. 3.3   (a) SUMFILTER block representation in FPAA software tool and, (b) its switched-capacitor circuit structure.

Fig. 3.3 shows an unconnected SUMFILTER block used in an FPAA device's software tool, and the device's switched-capacitor circuit structure. The CAM creates a summing stage with a maximum of three inputs and includes a single pole low pass filter. The inputs can be either inverting or non-inverting so that both sums and differences can be created in the transfer function. Each sampled input branch has a programmable gain. The sum of the input voltages is run through the single pole low pass filter which has a programmable corner frequency to produce a valid continuous output. The transfer function of this CAM is defined by the following equation [5]:

$$V_{out}(s) = \frac{2\pi f_0 \left[\pm G_1 V_{input1}(s) \pm G_2 V_{input2}(s) \pm G_3 V_{input3}(s)\right]}{s + 2\pi f_0} \quad (3.1)$$

The numbered *(G)* variables are the GAINs of the various input branches, and the numbered V_{Input} variables are the input voltages at the various input branches. The third term of this equation will only be activated if the corresponding CAM option input is turned on. The sign of each term is dependent on the polarity selected for each input branch in the CAM options. The terms are added for non-inverting inputs and subtracted for inverting inputs. The capacitor values are chosen based on the best ratios of the capacitors that satisfy the following relations [5]:

$$f_0 = \frac{f_c}{\pi} \cdot \frac{C_{out}}{2C_{int} + C_{out}}$$

$$G_1 = \frac{C_{inA}}{C_{out}}, \quad G_2 = \frac{C_{inB}}{C_{out}}, \quad G_3 = \frac{C_{inC}}{C_{out}} \quad (3.2)$$

Fig. 3.4 Experimental setup for FPAA-based implementations.

After the system implementation is modeled in the FPAA software tool, these models are downloaded to the FPAA development board via the serial interface. Experimental measurements can be obtained from the I/O connections on the FPAA board. A typical experimental setup for the AN221K04 FPAA board is shown in Fig. 3.4. This system includes a computer, a FPAA development board and a PC-compatible virtual measurement system using a PC oscilloscope module. The PC oscilloscope module incorporates a software that transitions as an oscilloscope and spectrum analyzer. This system is flexible and easy to use, and has many advantages over conventional instrumentation.

3.2 FPAA-Based Implementations of Chua's Circuit Model

As stated in the former chapters, it has been verified that Chua's circuit model can be realized with many different nonlinear functions including piecewise-linear function, cubic-like function, piecewise-quadratic function and some trigonometric functions. The implementation and experimental investigation of Chua's circuit modeled with different nonlinear functions require a significant variety of circuit hardware. Several circuit blocks, such as negative impedance converters via op amps, analog multipliers and full-wave rectifiers have been used for realizing the referred nonlinear functions. FPAA can be effectively used instead of these implementations of Chua's circuit. This programmable device is more efficient, simple and economical than using individual op-amps, comparators, analog multipliers and other discrete components. Nonlinear function blocks used in Chua's circuit can be modeled with FPAA programming, and a model can be rapidly changed for realizing other nonlinear functions.

In the next section, we introduce FPAA-based Chua's circuit implementations designed with respect to different nonlinear functions [84]. Before FPAA-based modeling of Chua's circuit with four nonlinear functions, these models have been tested with numerical simulations. Because the FPAA device has a ±2V saturation level, rescaling may be required for some of the models with respect to simulation results. Because it has similar structural design mechanism to FPAA software, we propose the use of SIMULINK for numerical modeling before FPAA modeling. Here, we present four FPAA-based system implementations of Chua's circuit associated with four different nonlinear functions studied via numerical simulations in the former chapter.

3.2.1 *FPAA-based Chua's circuit model-I*

The first FPAA-based implementation of Chua's circuit uses a piecewise-linear nonlinear function, $f(x) = bx + 0.5(a-b)(|x+c| - |x-c|)$. Rescaled and arranged, the system for FPAA design is defined by the following equations:

$$\dot{x} = 10y - 3.2x + 2.95(|x+1| - |x-1|)$$

$$\dot{y} = x - y + z \tag{3.3}$$

$$\dot{z} = -14.87y$$

The FPAA implementation scheme is shown in Fig. 3.5.

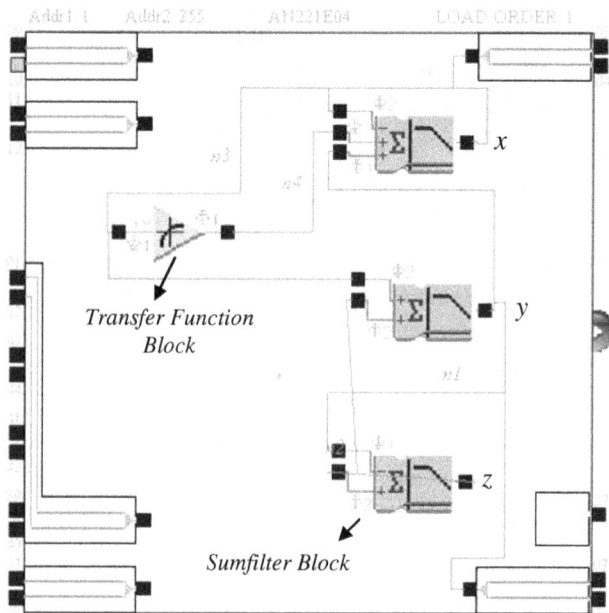

Fig. 3.5 FPAA implementation scheme of Chua's circuit with piecewise-linear function.

The blocks used in the implementation have been described in Table 3.2 with parameter settings.

Table 3.2 FPAA modules and parameter settings for the first Chua's circuit implementation defined by Eq. (3.3).

Name	Options		Parameters	
SumFilter1 (SumFilter v1.1.2)	Output Changes On	*Phase 2*	Corner Frequency [kHz]	*0.799*
	Input 1	*Inverting*	Gain 1 (UpperInput)	*2.20*
	Input 2	*Non-inverting*	Gain 2 (MiddleInput)	*2.95*
Anadigm (Approved)	Input 3	*Non-inverting*	Gain 3 (LowerInput)	*10.0*
SumFilter2 (SumFilter v1.1.2)	Output Changes On	*Phase 1*	Corner Frequency [kHz]	*0.800*
	Input 1	*Non-inverting*	Gain 1 (UpperInput)	*1.00*
	Input 2	*Non-inverting*	Gain 2 (LowerInput)	*1.00*
Anadigm (Approved)	Input 3	*Off*		
SumFilter3 (SumFilter v1.1.2)	Output Changes On	*Phase 1*	Corner Frequency [kHz]	*0.799*
	Input 1	*Inverting*	Gain 1 (UpperInput)	*9.90*
	Input 2	*Non-inverting*	Gain 2 (LowerInput)	*1.00*
Anadigm (Approved)	Input 3	*Off*		
TransferFunction1 (TransferFunction v1.1.0)	Output Hold	*Off*		
	Input Full Scale	*3 Volts*		
Anadigm (Approved)				

As shown in the FPAA implementation scheme in Fig.3.5, system state-variables *x, y* and *z* are represented at the outputs of SUMFILTER blocks. Circuit gains are implemented by SUMFILTER block gains. A user-defined TRANSFER FUNCTION block was used for implementing piecewise-linear nonlinear function. This transfer function module produces an output voltage with 256 quantization steps according to a lookup table constituted by the user. The chaotic dynamics and double-scroll chaotic attractor obtained from the first FPAA-based Chua's circuit using piecewise-linear nonlinearity are shown respectively in Fig. 3.6.

3.2.2 *FPAA-based Chua's circuit model-II*

After completing an experiment on the FPAA-based model and getting results, the same FPAA device can be reconfigured for another modeling on the fly. So, in our modeling and experimental study for Chua's circuit, we changed the nonlinear function used in the first implementation by keeping the common part of the circuit. The second FPAA-based implementation uses a cubic-like nonlinear function and this model is defined by the following state equations:

$$\dot{x} = 2.4y + 2.7x - 3.925x^3$$

$$\dot{y} = 4.167\,x - y + 7.083\,z \qquad (3.4)$$

$$\dot{z} = -2.099\,y$$

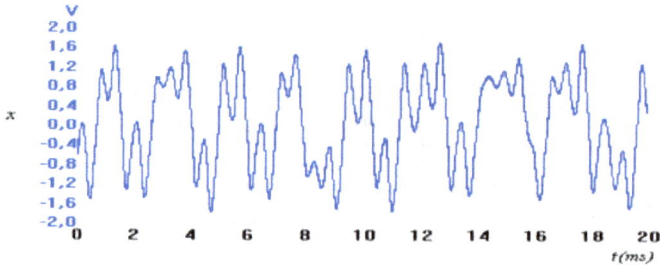

(a)

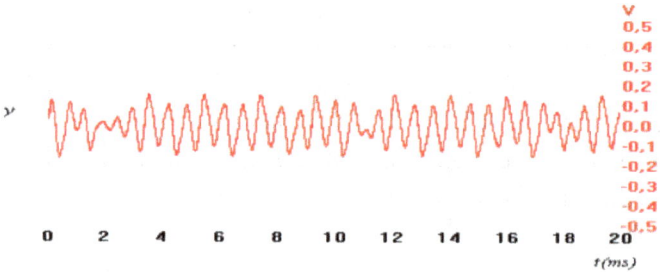

(b)

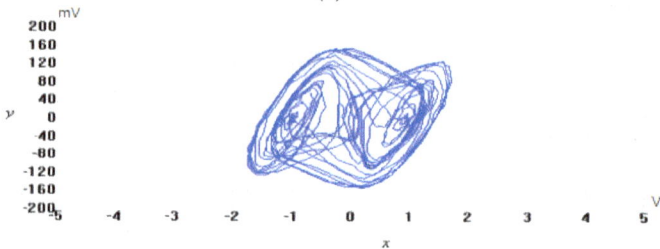

(c)

Fig. 3.6 Experimental results obtained from the first FPAA-based implementation of Chua's circuit, (a)-(b) time responses of *x(t)* and *y(t)* chaotic dynamics, (c) chaotic attractor projection in *x-y* plane.

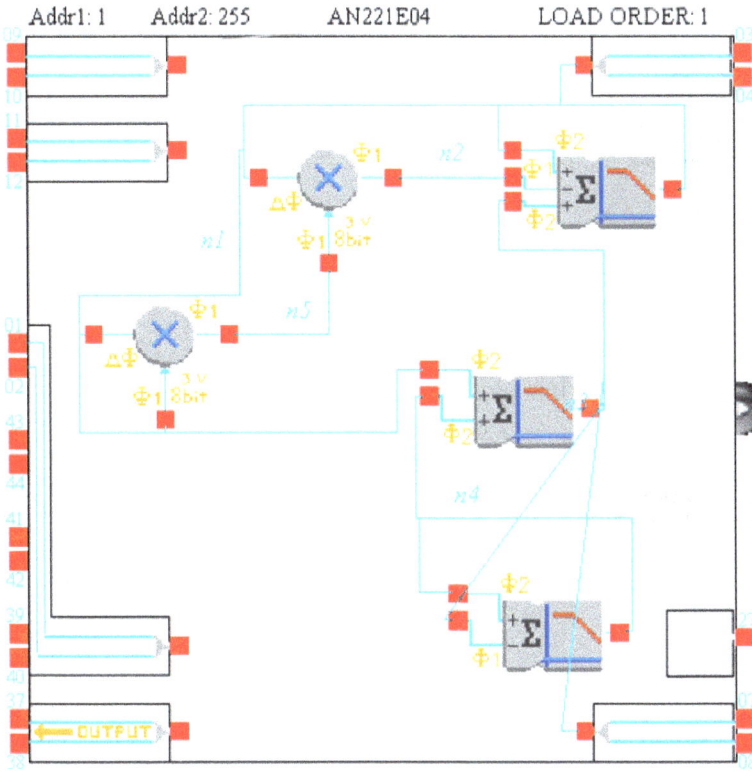

Fig. 3.7 FPAA implementation scheme of Chua's circuit with cubic-like function.

The FPAA implementation scheme of the second model is shown in Fig. 3.7. The blocks used in the implementation have been described in Table 3.3 with parameter settings.

In addition to common SUMFILTER blocks, MULTIPLIER blocks were used to implement the cubic-like nonlinear function. After the reconfiguration and downloading process, the experimental measurements were realized as in the first one. The chaotic dynamics and the double-scroll chaotic attractor illustrations belonging to this modeling are shown in Fig. 3.8.

A Practical Guide for Studying Chua's Circuits

Table 3.3 FPAA modules and parameter settings for the second Chua's circuit
implementation defined by Eq. (3.4).

Name	Options		Parameters	
SumFilter1 (SumFilter v1.1.2) *Anadigm (Approved)*	Output Changes On	*Phase 1*	Corner Frequency [kHz]	*0.400*
	Input 1	*Non-inverting*	Gain 1 (UpperInput)	*3.70*
	Input 2	*Inverting*	Gain 2 (MiddleInput)	*3.90*
	Input 3	*Non-inverting*	Gain 3 (LowerInput)	*2.40*
SumFilter2 (SumFilter v1.1.2) *Anadigm (Approved)*	Output Changes On	*Phase 1*	Corner Frequency [kHz]	*0.399*
	Input 1	*Non-inverting*	Gain 1 (UpperInput)	*4.17*
	Input 2	*Non-inverting*	Gain 2 (LowerInput)	*7.08*
	Input 3	*Off*		
SumFilter3 (SumFilter v1.1.2) *Anadigm (Approved)*	Output Changes On	*Phase 1*	Corner Frequency [kHz]	*0.400*
	Input 1	*Non-inverting*	Gain 1 (UpperInput)	*1.00*
	Input 2	*Inverting*	Gain 2 (LowerInput)	*2.10*
	Input 3	*Off*		
Multiplier1 (Multiplier v1.2.0) *Anadigm (Approved)*	Sample and Hold	*Off*	Multiplication Factor	*1.00*
	Y Input Full Scale	*3 Volts*		
Multiplier2 (Multiplier v1.2.0) *Anadigm (Approved)*	Sample and Hold	*Off*	Multiplication Factor	*1.00*
	Y Input Full Scale	*3 Volts*		

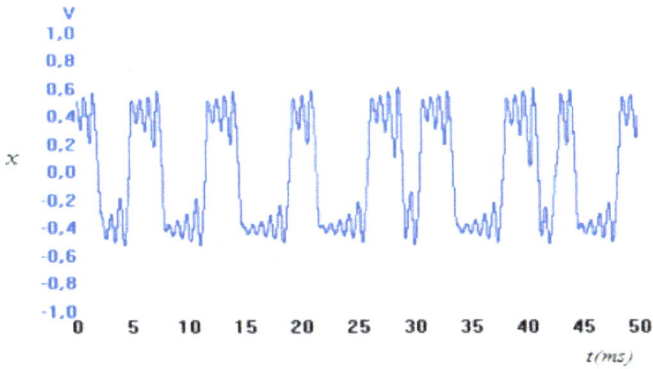

(a)

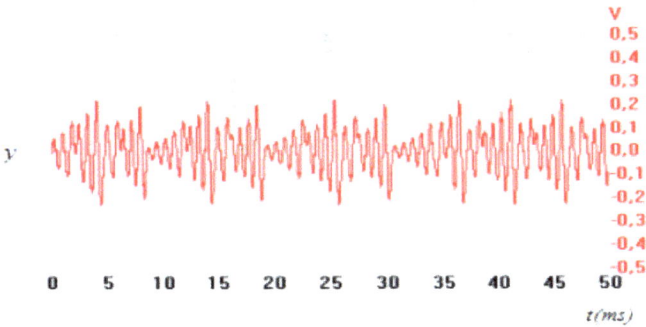

(b)

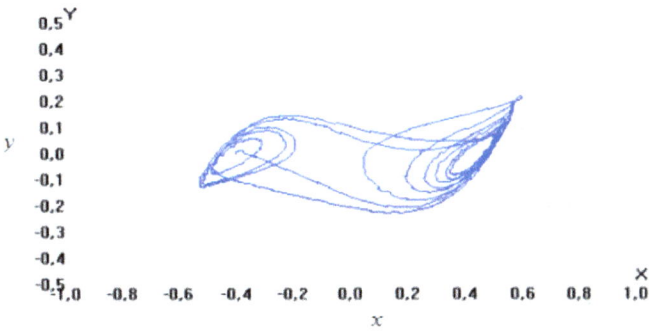

(c)

Fig. 3.8 Experimental results obtained from the second FPAA implementation of Chua's circuit, (a)–(b) time responses of $x(t)$ and $y(t)$ chaotic dynamics, (c) chaotic attractor projection in x-y plane.

3.2.3 *FPAA-based Chua's circuit model-III*

The third FPAA-based implementation of Chua's circuit, shown in Fig. 3.9, has been constructed in the FPAA environment according to the model defined by Eq. (3.5):

$$\dot{x} = 6.3y - 6.3x + 20\tanh(0.38x)$$

$$\dot{y} = 0.63x - 0.63y + z \qquad\qquad (3.5)$$

$$\dot{z} = -5.6y$$

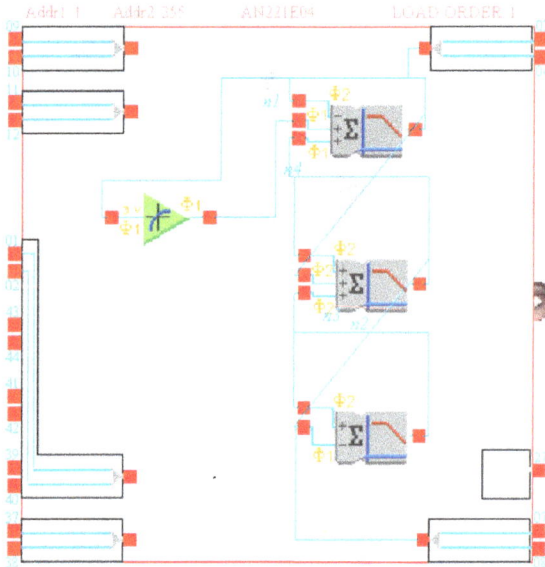

Fig. 3.9 FPAA implementation scheme of Chua's circuit with a hyperbolic nonlinear function.

This implementation uses a hyperbolic nonlinear function including *tanh(.)* term. The blocks used in this implementation have been described in Table 3.4 with parameter settings. In addition to common SUMFILTER blocks used in other models, a user-defined TRANSFER FUNCTION block was used for implementing hyperbolic function including *tanh(.)* term. After the reconfiguration and downloading process, the experimental measurements are obtained as in the other cases. The chaotic dynamics and the double-scroll chaotic attractor illustrations are shown in Fig. 3.10.

Table 3.4 FPAA modules and parameter settings for the third Chua's circuit implementation defined by Eq. (3.5).

Name	Options		Parameters	
SumFilter1 (SumFilter v1.1.2) *Anadigm (Approved)*	Output Changes On	*Phase 2*	Corner Frequency [kHz]	0.400
	Input 1	*Inverting*	Gain 1 (UpperInput)	5.30
	Input 2	*Non-inverting*	Gain 2 (MiddleInput)	30.0
	Input 3	*Non-inverting*	Gain 3 (LowerInput)	6.30
SumFilter2 (SumFilter v1.1.2) *Anadigm (Approved)*	Output Changes On	*Phase 1*	Corner Frequency [kHz]	0.401
	Input 1	*Non-inverting*	Gain 1 (UpperInput)	0.364
	Input 2	*Non-inverting*	Gain 2 (MiddleInput)	0.909
	Input 3	*Non-inverting*	Gain 3 (LowerInput)	1.00
SumFilter3 (SumFilter v1.1.2) *Anadigm (Approved)*	Output Changes On	*Phase 1*	Corner Frequency [kHz]	0.400
	Input 1	*Non-inverting*	Gain 1 (UpperInput)	1.00
	Input 2	*Inverting*	Gain 2 (LowerInput)	5.60
	Input 3	*Off*		
TransferFunction1 (TransferFunction v1.1.0) *Anadigm (Approved)*	Output Hold	*Off*		
	Input Full Scale	*3 Volts*		

3.2.4 *FPAA-based Chua's circuit model-IV*

The last FPAA-based implementation of Chua's circuit model in this chapter uses a piecewise-quadratic nonlinear function as shown in Fig. 3.11. This model is defined by the following state equations:

$$\dot{x} = 3.03y + 1.428x - 0.635x|3.3x|$$

$$\dot{y} = 3.3x - y + 4z \qquad\qquad (3.6)$$

$$\dot{z} = -3.7175y$$

(a)

(b)

(c)

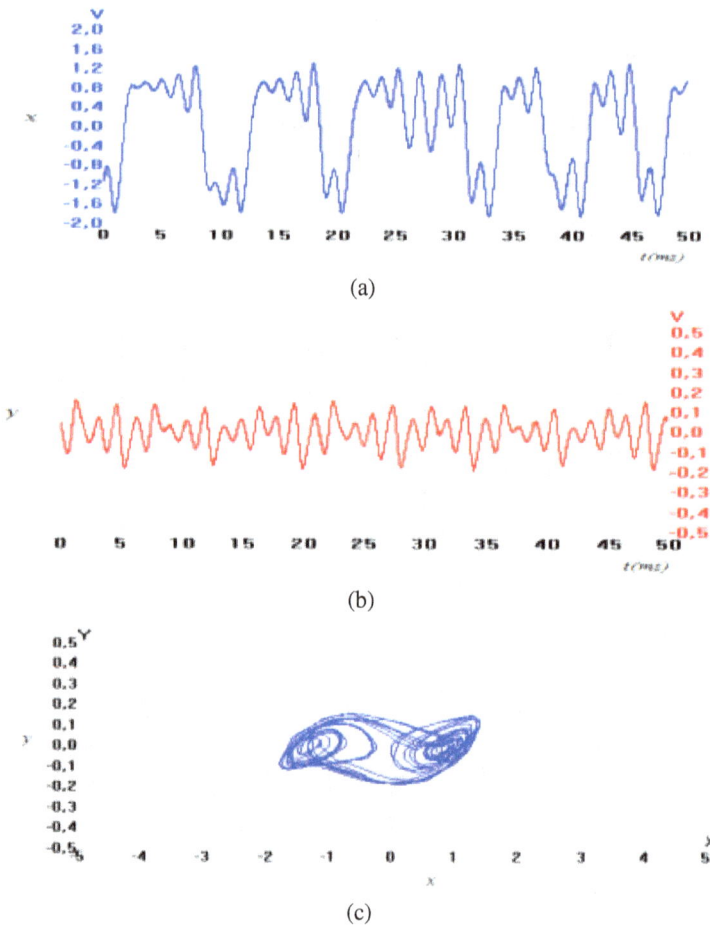

Fig. 3.10 Experimental results obtained from the third FPAA implementation of Chua's circuit, (a)–(b) time responses of $x(t)$ and $y(t)$ chaotic dynamics, (c) chaotic attractor projection in x-y plane.

The blocks used in the implementation have been described in Table 3.5 with parameter settings. System state-variables x, y and z are represented at the output of SUMFILTER blocks as in the former implementations.

Circuit gains are implemented by SUMFILTER block gains. MULTIPLIER and RECTIFIER blocks were used for implementing the piecewise-quadratic nonlinear function.

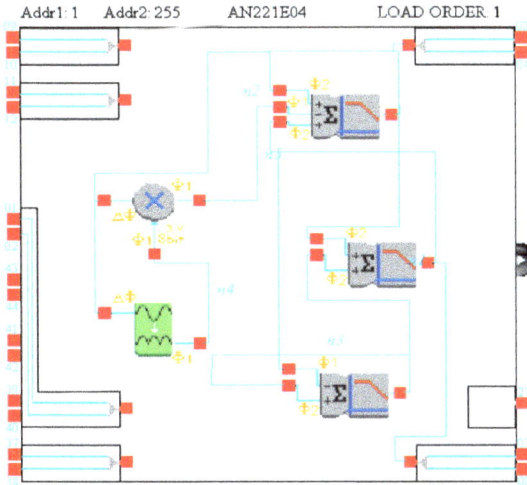

Fig. 3.11   FPAA implementation scheme of Chua's circuit with piecewise-quadratic nonlinear function.

Table 3.5 FPAA modules and parameter settings for the fourth Chua's circuit implementation defined by Eq. (3.6).

Name	Options		Parameters	
SumFilter1 (SumFilter v1.1.2) Anadigm (Approved)	Output Changes On	Phase 1	Corner Frequency [kHz]	0.400
	Input 1	Non-inverting	Gain 1 (UpperInput)	2.46
	Input 2	Inverting	Gain 2 (MiddleInput)	0.692
	Input 3	Non-inverting	Gain 3 (LowerInput)	3.54
SumFilter2 (SumFilter v1.1.2) Anadigm (Approved)	Output Changes On	Phase 1	Corner Frequency [kHz]	0.400
	Input 1	Non-inverting	Gain 1 (UpperInput)	3.30
	Input 2	Non-inverting	Gain 2 (LowerInput)	4.00
	Input 3	Off		
SumFilter3 (SumFilter v1.1.2) Anadigm (Approved)	Output Changes On	Phase 1	Corner Frequency [kHz]	0.399
	Input 1	Inverting	Gain 1 (UpperInput)	3.71
	Input 2	Non-inverting	Gain 2 (LowerInput)	1.00
	Input 3	Off		
Multiplier1 (Multiplier v1.2.0) Anadigm (Approved)	Sample and Hold	Off	Multiplication Factor	1.00
	Y Input Full Scale	3 Volts		
RectifierHalf1 (RectifierHalf v2.1.5) Anadigm (Approved)	Rectification	Full Wave	Gain	3.30
	Polarity	Non-inverting		
	Output Phase	Phase 1		

The piecewise-quadratic nonlinear function could also be implemented using a user-defined TRANSFER FUNCTION block. To show the design flexibility of FPAA programming, we prefer to use an alternative approach.

The chaotic dynamics and double-scroll chaotic attractor obtained from the last FPAA-based chaotic system using piecewise-quadratic nonlinearity are shown respectively in Fig. 3.12.

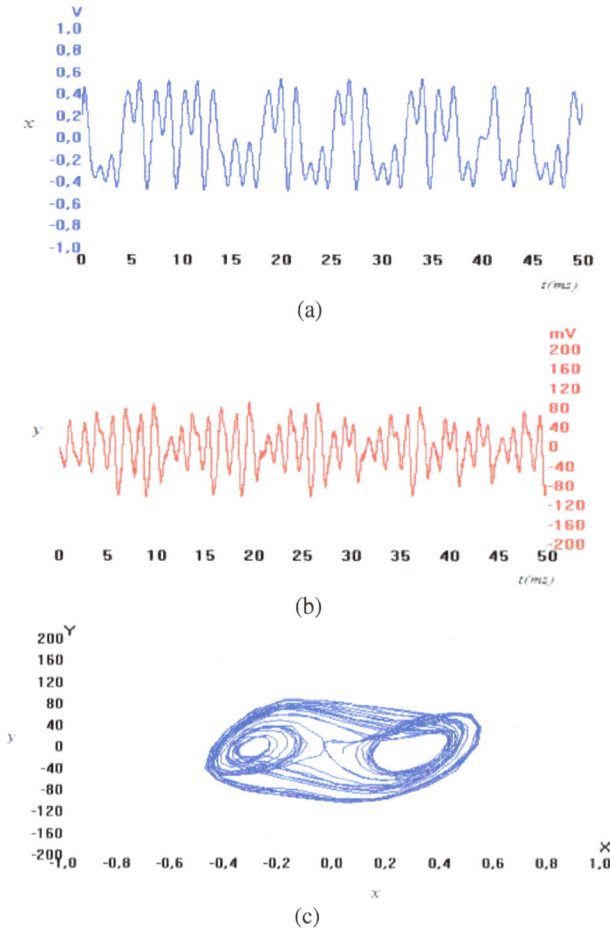

(a)

(b)

(c)

Fig. 3.12 Experimental results for the fourth FPAA implementation of Chua's circuit, (a)–(b) time responses of x(t) and y(t) chaotic dynamics, (c) chaotic attractor projection in x-y plane.

Chapter 4

Mixed-Mode Chaotic Circuit (MMCC): A Versatile Chaotic Circuit Utilizing Autonomous and Nonautonomous Chua's Circuits

In this chapter, we describe an interesting switched chaotic circuit using autonomous and nonautonomous Chua's circuits. We first consider the original design of the mixed-mode chaotic circuit (MMCC). Further, we consider alternative circuit implementations of the proposed circuit.

Chaotic circuits for use in chaotic secure communication systems must not only have a simple design but also a structure that is able to provide greater reliability in the form of a wide range of parameter variations and extra security keys. For this purpose, we have designed a mixed-mode chaotic circuit, which has both autonomous and nonautonomous chaotic dynamics, using a switching method. In this design, an autonomous Chua's circuit [24] and a nonautonomous Murali-Lakshmanan-Chua (MLC) circuit [101] were combined through a common dynamic using a switching method. In this way, the mixed-mode circuit operates either in the chaotic regime determined by the autonomous circuit part or in the chaotic regime determined by the nonautonomous circuit part, depending on the state of the switches.

4.1 Design Procedure of Mixed-Mode Chaotic Circuit

The primary design of the mixed-mode chaotic circuit [68] is shown in Fig. 4.1. In this design, an autonomous Chua's circuit and a nonautonomous MLC circuit were combined via two voltage buffers and switches. In this way, depending on the states of the switches and directions of the voltage buffers, this mixed-mode circuit operates either in the chaotic regime determined by the autonomous circuit part or in the chaotic regime determined by the nonautonomous circuit part. Because two nonlinear resistors and two voltage buffers are used in this design, the circuit structure is very complex. Therefore, we simplified the circuit design. To simplify the circuit design in Fig. 4.1, we proposed the circuit in Fig. 4.2.

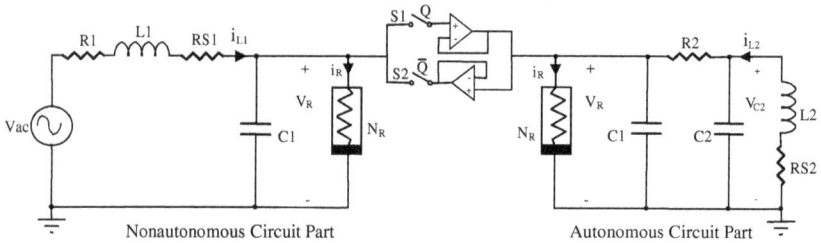

Fig. 4.1 The primary design of the mixed-mode chaotic circuit.

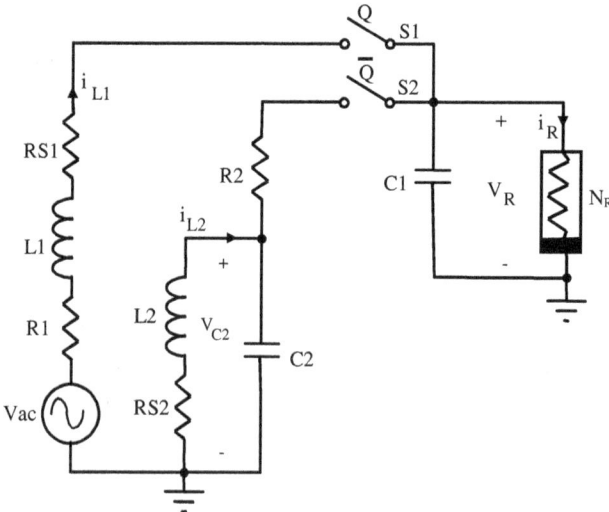

Fig. 4.2 Mixed-mode chaotic circuit.

In this second circuit design [69, 73], nonlinear resistor N_R and capacitor C_1 in Fig. 4.1 were chosen as the common circuit elements. We first show and verify the circuit's static performance using manual switching. When the states of switches are S1-ON and S2-OFF in Fig. 4.2, we have the standard nonautonomous MLC circuit exhibiting a double-scroll chaotic attractor. In this case, the circuit is represented by the following set of two first-order nonautonomous differential equations:

$$C_1 \frac{dV_R}{dt} = i_{L1} - f(V_R)$$

$$L_1 \frac{di_{L1}}{dt} = -i_{L1}(R_1 + R_{S1}) - V_R + A\sin(wt)$$

(4.1)

where (A) is the amplitude and (w) is the angular frequency of the external periodic voltage source V_{ac} in Fig. 4.2. For this nonautonomous circuit part, we fixed the parameters of the circuit elements as $C_1 = 10$ nF, $L_1 = 18$ mH, $R_1 = 1340\ \Omega$, $R_{S1} = 12.5\ \Omega$ and $A = 0.15$ V. Frequency of the external forcing source is 8890 Hz, for which the circuit generates double-scroll chaotic oscillations. The amplitude of the external forcing source can be used as the bifurcation parameter. By increasing the amplitude (A) from zero upwards, the circuit exhibits the complex dynamics of bifurcation and chaos. The circuit's chaotic waveform and double-scroll chaotic attractor are shown in Fig. 4.3(a) and (b), respectively.

When the states of switches are S1-OFF and S2-ON in Fig. 4.2, we have the standard autonomous Chua's circuit exhibiting a double-scroll Chua's chaotic attractor. In this case, the circuit is described by the following set of three first-order autonomous differential equations:

$$L_2 \frac{di_{L2}}{dt} = -V_{C2} - i_{L2} \cdot R_{S2}$$

$$C_2 \frac{dV_{C2}}{dt} = i_{L2} - \frac{1}{R_2}(V_{C2} - V_R)$$

$$C_1 \frac{dV_R}{dt} = \frac{1}{R_2}(V_{C2} - V_R) - f(V_R)$$

(4.2)

By reducing the resistor R_2 in Fig. 4.2 from 2000 Ω towards zero, this circuit part exhibits complex dynamics of bifurcation and chaos.

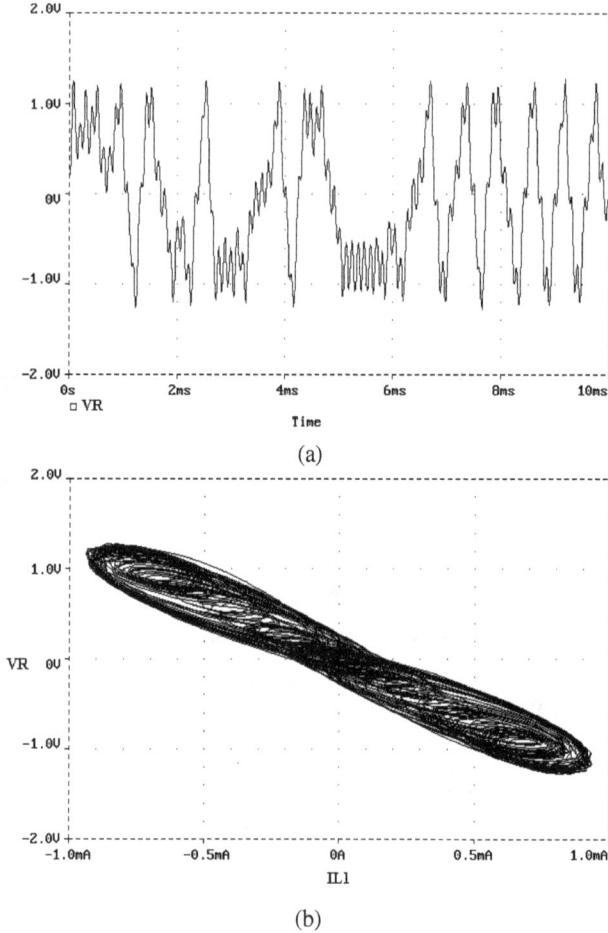

(a)

(b)

Fig. 4.3   The chaotic waveform and double-scroll chaotic attractor observed in the nonautonomous mode of the MMCC.

We fixed the parameters of the circuit elements as C_1 = 10 nF, C_2 = 100 nF, L_2 = 18 mH, R_2 = 1700 Ω and R_{S2} = 12.5 Ω, for which the circuit exhibits a double-scroll chaotic attractor. The circuit's chaotic waveform and double-scroll chaotic attractor are shown in Fig. 4.4(a)-(b), respectively.

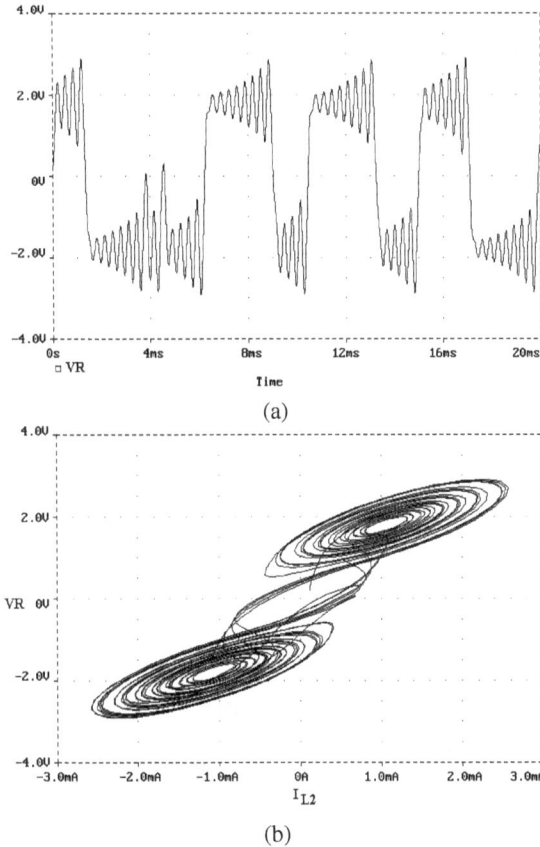

(a)

(b)

Fig. 4.4 The chaotic waveform and double-scroll chaotic attractor observed in the autonomous mode of the MMCC.

In order to evaluate the dynamic performance of the mixed-mode chaotic circuit, we apply two complementary square waves, Q and $\overline{Q}$, to the gates of the analog switches S1 and S2 in Fig. 4.2 so that the switching operation is going on continuously. Impulse width and period of each square wave were chosen as 5 ms and 10 ms. Mixed-mode chaotic waveform and double-scroll chaotic attractor of the proposed circuit are illustrated in Fig. 4.5(a) and (b).

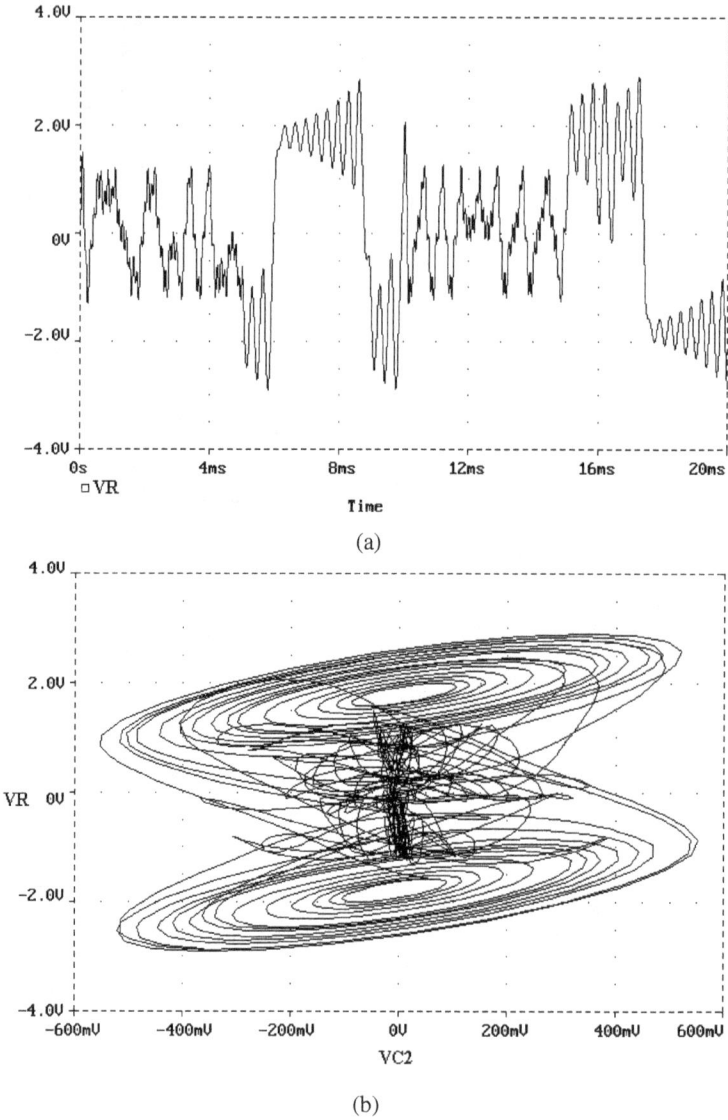

(a)

(b)

Fig. 4.5 Mixed-mode chaotic waveform and double-scroll chaotic attractor.

Instead of a periodic digital signal, the various node voltages in the mixed-mode chaotic circuit can be used to control the switches as shown in Fig. 4.6. Here, node-1 was chosen as the control node. While switch S1 is driven by node-1, switch S2 is driven by the complement of node-1.

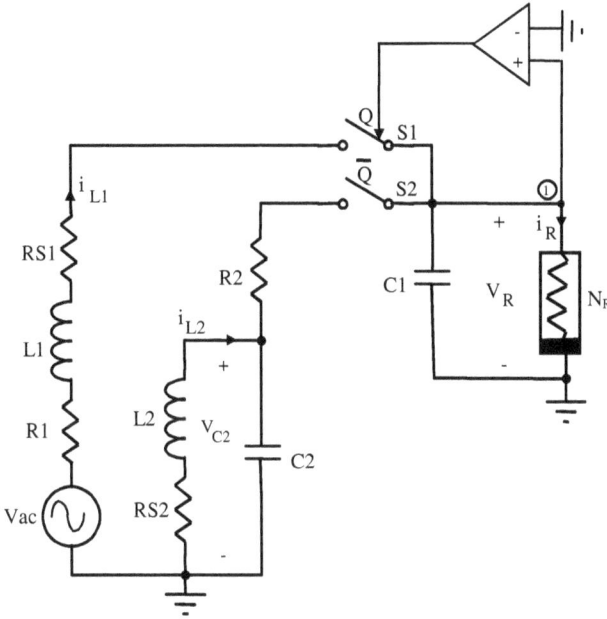

Fig. 4.6 Mixed-mode chaotic circuit with node voltage controlled switching method.

4.2 Improved Realizations of the MMCC

In this section, we discuss some alternative implementations of the MMCC. Because the MMCC consists of autonomous and nonautonomous Chua's chaotic circuits, the circuit topologies proposed for nonlinear resistor and inductor elements in Chua's circuit can be applied to the MMCC for improved and/or alternative realization of it.

4.2.1 *FTFN-based MMCC*

An FTFN-based MMCC [74] is shown in Fig. 4.7. In this realization of the mixed-mode chaotic circuit, as a major improvement, FTFN-based inductance simulators are used instead of the floating inductance L_1 and grounded inductance L_2 in Fig. 4.2.

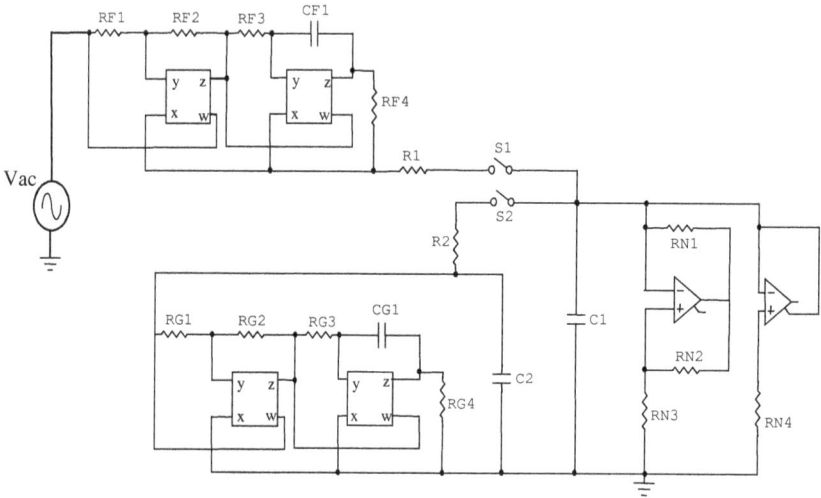

Fig. 4.7 An FTFN-based MMCC circuit.

An FTFN-based floating inductance simulator was introduced in Chapter 1. In Fig. 4.7, the following element values are chosen: $R_3 = R_4 = R_5 = R_6 = 1$ KΩ, $C_4 = 18$ nF to simulate $L_1 = L_2 = 18$ mH. Floating inductance is also used as grounded inductor by connecting one port of the floating inductance to the ground for simplicity. The PSPICE simulations were performed using a CMOS realization of FTFN, shown in Fig. 1.14(a) of the first chapter.

In this implementation, we used the CFOA-based circuit structure for nonlinear resistor described in Fig. 1.11 of the first chapter. The CFOA-based circuit realization of nonlinear resistor combines attractive features of the current feedback op-amp operating in both voltage and current modes. The simulation results in Fig. 4.8 and Fig. 4.9 depict the autonomous and nonautonomous chaotic dynamics of the MMCC. These results verify the FTFN-based circuit design considerations.

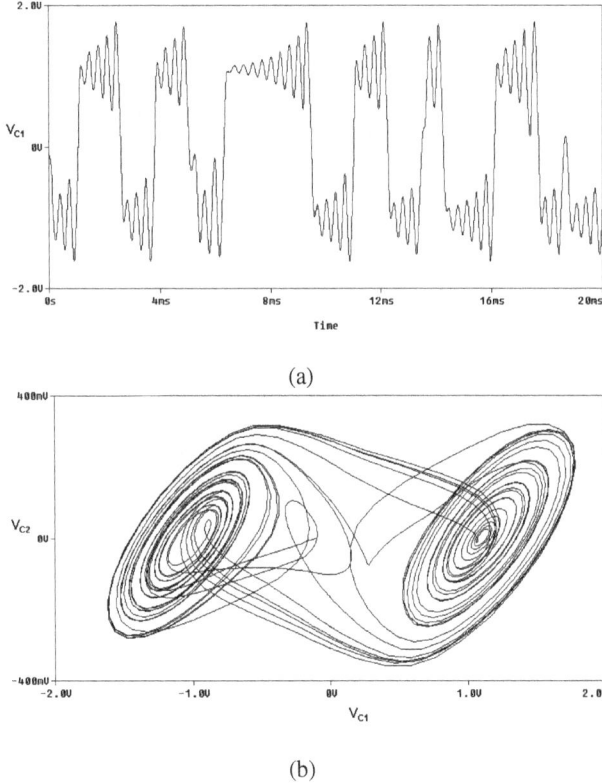

(a)

(b)

Fig. 4.8 When improved mixed-mode chaotic circuit of Fig.4.7 oscillates in autonomous mode (S1-OFF, S2-ON), (a) the chaotic waveform of the voltage across capacitor C_1, and (b) double-scroll chaotic attractor.

4.2.2 *CFOA-based MMCC*

In this section, we present a versatile practical implementation of the MMCC [80]. Fig. 4.10 depicts the experimental realization of a CFOA-based inductorless MMCC formed by CFOA-based Chua's diode and CFOA-based floating and grounded inductance simulators.

(a)

(b)

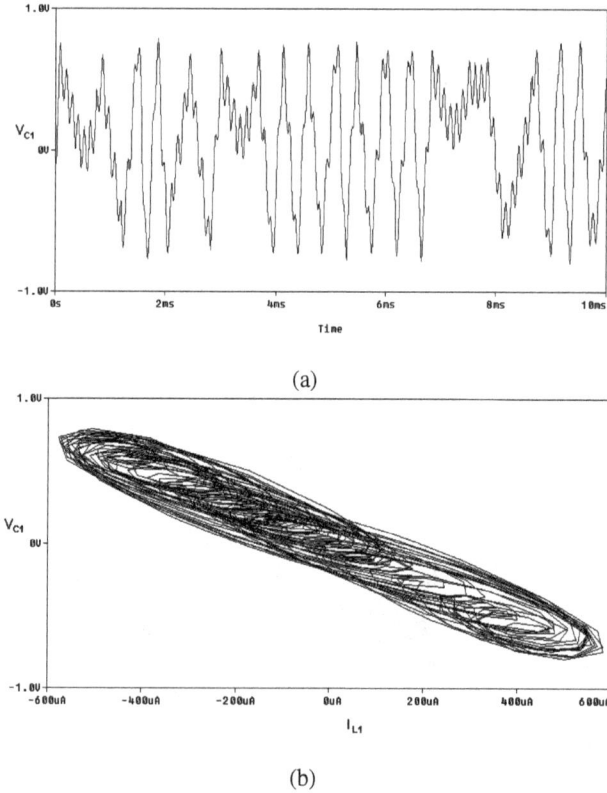

Fig. 4.9   When improved mixed-mode chaotic circuit of Fig. 4.7 oscillates in nonautonomous mode (S1-ON, S2-OFF), (a) the chaotic waveform of the voltage across capacitor C_1, and (b) double-scroll chaotic attractor.

In this experimental setup, AD844 ICs have been used as the current feedback amplifier. AD844 type CFOA provides significant advantages over the voltage-mode operational amplifier (VOA) for higher-frequency operation. AD844 IC, which has current feedback architecture, is a high-speed operation amplifier, and it combines high bandwidth and very fast large signal response with excellent dc performance. AD844 IC provides a closed-loop bandwidth (up to 60 MHz), which is almost independent of the closed-loop gain. Its slew rate is typically around 2000V/μs. Hence, this realization using AD844ICs is free from the slew rate limitations of traditional voltage mode operational amplifiers.

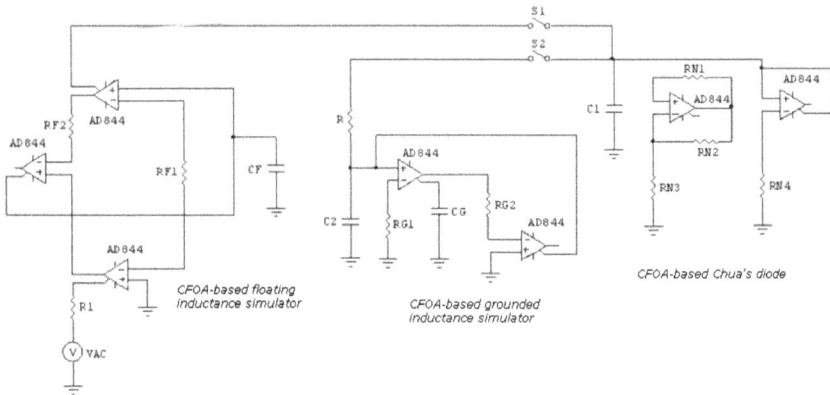

Fig. 4.10 CFOA-based mixed-mode chaotic circuit.

Due to the use of CFOAs for floating and grounded inductance simulators, and Chua's diode, all state variables are made available in a direct manner. Also, a buffered output is available.

In the experimental setup, Chua's diode has been implemented by using two AD844 type CFOAs biased with ±9 V and four resistors, $R_{N1} = R_{N2} = 22$ kΩ, $R_{N3} = 500$ Ω and $R_{N4} = 2.2$ kΩ. While the floating inductance simulator block is configured by three AD844 type CFOAs biased with ±9 V, two resistors $R_{F1} = R_{F2} = 1$ kΩ and one capacitor $C_F = 18$ nF to simulate $L_1 = 18$ mH, a grounded inductance simulator block is configured by two AD844 type CFOAs biased with ±9 V, two resistors $R_{G1} = R_{G2} = 1$ kΩ and one capacitor $C_G = 18$ nF to simulate $L_2 = 18$ mH. Routine analysis for CFOA-based floating and grounded inductance simulators yields equivalent inductance as

$$L_{eq} = R_{F1} \cdot R_{F2} \cdot C_F = R_{G1} \cdot R_{G2} \cdot C_G \qquad (4.3)$$

The circuit element values for inductance simulator configurations are determined according to Eq. (4.3). In addition to the above arrangements for implementing nonlinear resistor and inductance simulators, we fixed the other circuit parameters as $C_1 = 10$ nF, $C_2 = 100$ nF, R = 2 kΩ pot, $R_1 = 1340$ Ω, the frequency f = 8890 Hz and the amplitude of ac source V_{ac} A = 0.15 V. As the electronic switching device, the 4016 IC, which contains four voltage-controlled analog switches, was used in our experiments.

4.2.2.1 *Experimental results*

In our experiments, we first verified the circuit's static performance using manual switching. By applying positive control voltage to S1 and negative control voltage to S2, we transformed the MMCC to a nonautonomous MLC circuit. In this operation mode, we measured the circuit's nonautonomous chaotic dynamics in time domain and x-y projection by using the digital storage oscilloscope interfaced with a computer. The nonautonomous chaotic dynamics and double-scroll chaotic attractor observed from the experiments are shown in Fig. 4. 11.

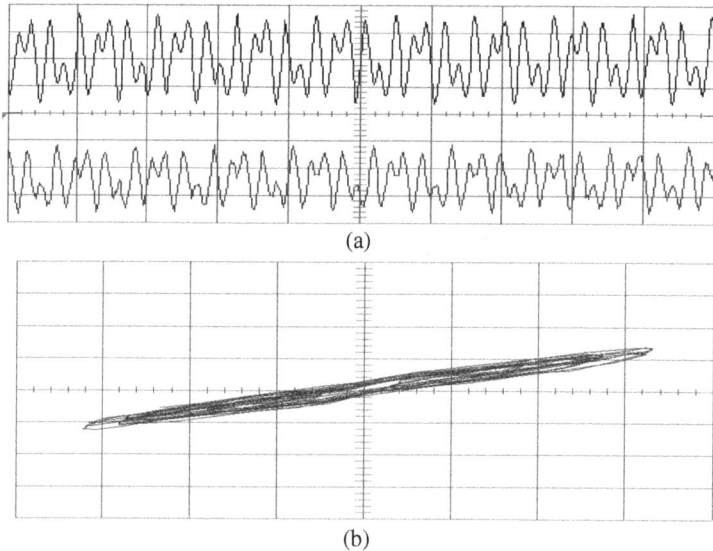

(a)

(b)

Fig. 4.11 (a) Chaotic circuit dynamics of CFOA-based mixed-mode chaotic circuit in nonautonomous operation mode; the upper trace V_{C1} (2V/div), the lower trace i_{L1} current output in the voltage form via a R_{sens}=100Ω current sensing resistor (500mV/div), time/div: 500μs/div, (b) The chaotic attractor observed in the V_{C1}-i_{L1} plane, x-axes:1V, y-axes: 500mV.

For monitoring i_{L1} current in voltage form at oscilloscope screen, we used a current sensing resistor, in the value of R_{sens} = 100 Ω. By changing the control signals, *i.e.*, in case of switch positions S1-OFF and S2-ON, we have an autonomous Chua's circuit. The chaotic dynamics and double-scroll Chua's attractor obtained from experimental measurements are given in Fig. 4.12. These results recorded in both operation modes confirm inductorless CFOA-based MMCC's static performance.

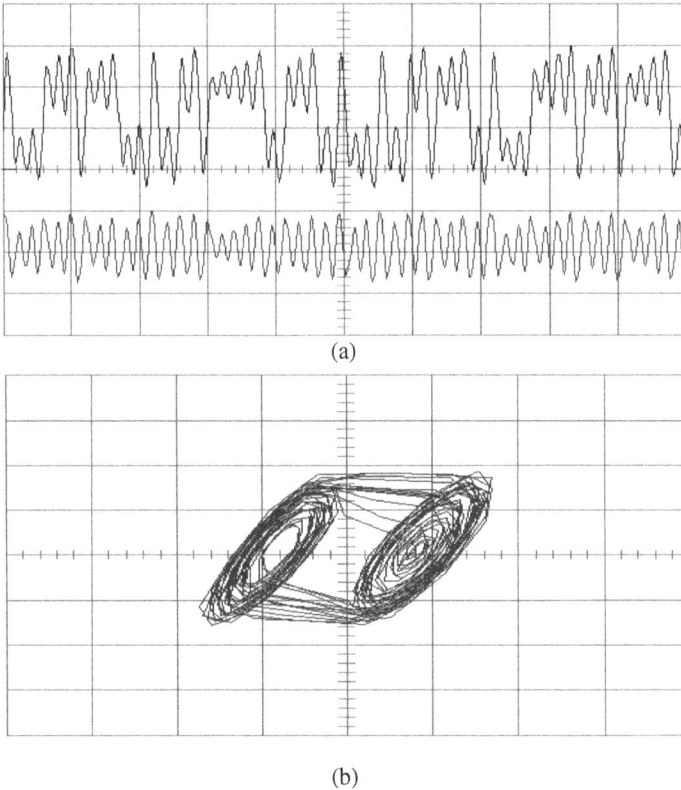

(a)

(b)

Fig. 4.12 (a) Chaotic circuit dynamics of CFOA-based mixed-mode chaotic circuit in autonomous operation mode; the upper trace V_{C1} (2 V/div), the lower trace V_{C2} (1 V/div), time/div: 2 ms/div, (b) chaotic attractor observed in the V_{C1}–V_{C2} plane, x-axes: 2 V, y-axes: 500 mV.

In order to verify dynamic performance of an inductorless CFOA-based MMCC, we applied two complementary square waves Q and $\overline{Q}$, which have 5 ms impulse width and 10 ms period, in our experiments. The experimental results including controlling signal, mixed-mode chaotic dynamic and chaotic attractor projection in V_{C2}–V_{C1} plane are illustrated in Fig. 4.13.

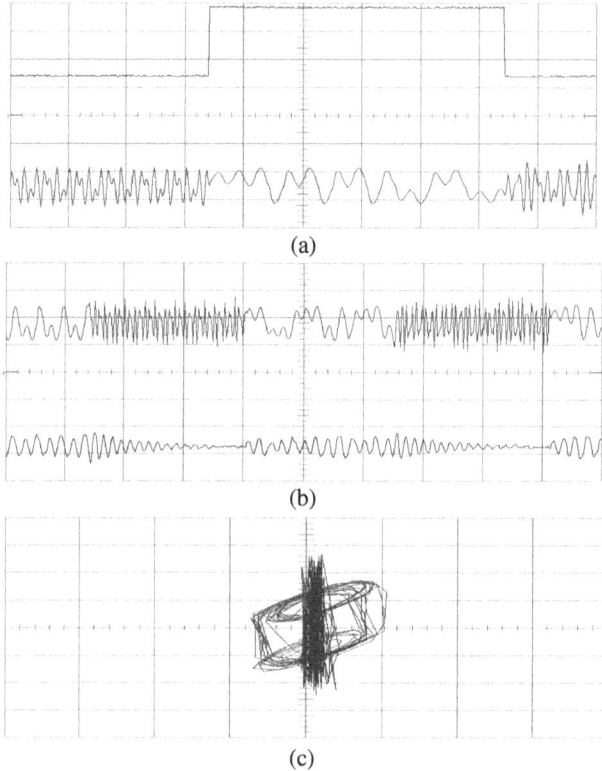

(a)

(b)

(c)

Fig. 4.13   (a) The transitions dependent on controlling signal in CFOA-based mixed-mode chaotic circuit; the upper trace controlling signal (5V/div), the lower trace V_{C1} (5 V/div), time/div: 1 ms/div, (b) mixed-mode chaotic dynamic of CFOA-based mixed-mode chaotic circuit; the upper trace V_{C1} (2 V/div), the lower trace V_{C2} (5 V/div), time/div: 2 ms/div, (c) chaotic attractor observed in the V_{C2}–V_{C1} plane, x-axes: 1 V, y-axes: 2 V.

To illustrate the switching mechanism in dynamic operation mode, different switching projections are shown in Fig. 4.14.

(a)

(b)

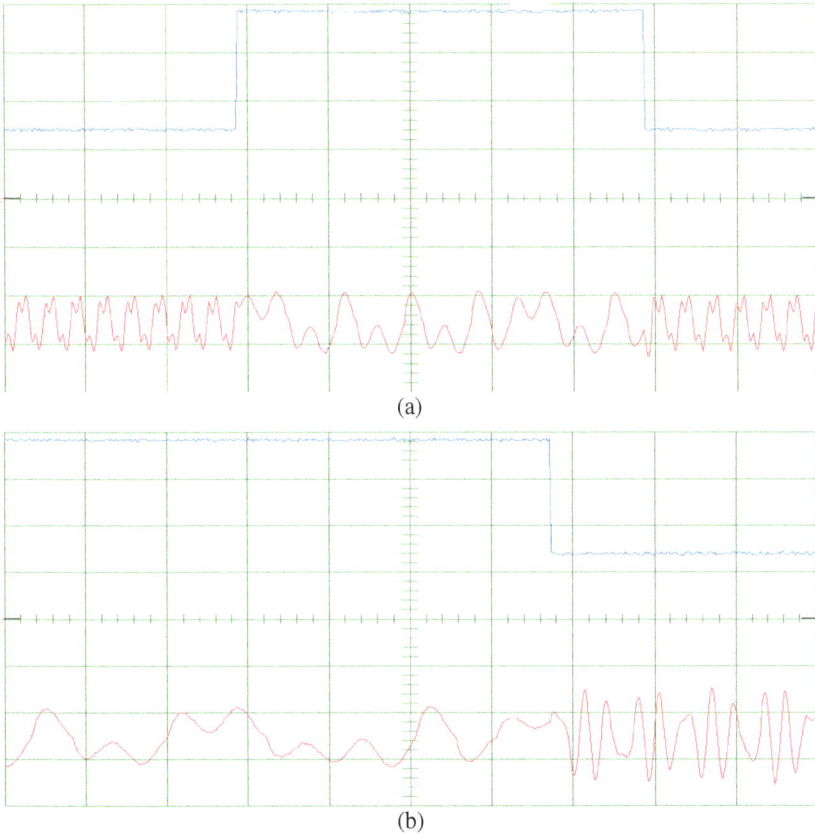

Fig. 4.14 Different switching projections in dynamic operation mode, (a) mixed-mode chaotic circuit combined with double-scroll autonomous dynamic and period-3 nonautonomous dynamic; the upper trace controlling signal (5 V/div), the lower trace V_{C1} (5 V/div), time/div: 1 ms/div, (b) transition of circuit dynamic from autonomous mode to nonautonomous mode; the upper trace controlling signal (5 V/div), the lower trace V_{C1} (5 V/div), time/div: 500 µs/div.

In order to demonstrate the proposed CFOA-based circuit's high-frequency performance, we examined the circuit by using an additional experimental configuration. In this configuration, we determined the circuit parameters in Fig. 4.10 as $C_1 = 100$ pF, $C_2 = 1$ nF, $C_F = C_G = 180$ pF, and frequency of ac voltage source V_{ac} f = 889 kHz by scaling down values of synthetic inductors and capacitors, and scaling up value of the frequency of ac voltage source V_{ac} by a factor of 100. The circuit's high-frequency autonomous chaotic dynamics are illustrated in Fig. 4.15(a).

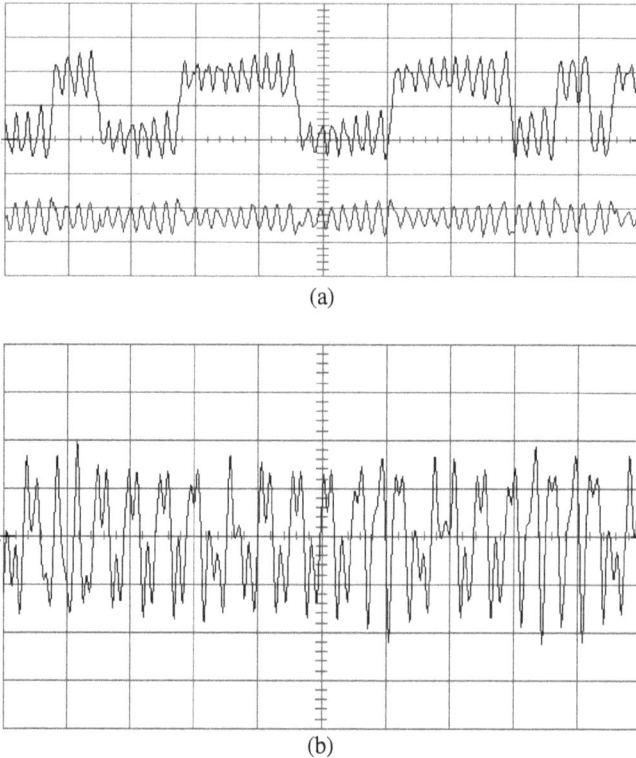

(a)

(b)

Fig. 4.15 For high-frequency experimental configuration of CFOA-based mixed-mode chaotic circuit, (a) high-frequency chaotic circuit dynamics observed in autonomous operation mode; the upper trace V_{C1} (500 mV/div), the lower trace V_{C2} (500 mV/div), time/div: 20 μs/div, (b) high-frequency chaotic circuit dynamic observed in nonautonomous operation mode, the trace V_{C1} (500 mV/div), time/div: 10 μs/div.

From laboratory experiments, we determined that the circuit's chaotic spectrum to be centered approximately on 300 kHz for this autonomous mode. The circuit's high-frequency nonautonomous chaotic dynamic is illustrated in Fig. 4.15(b). The circuit is able to exhibit nonautonomous chaotic dynamic in this nonautonomous mode, although a high-frequency (f = 889 kHz) ac voltage source is used. According to our experimental observations, the proposed circuit's chaotic frequency spectrum extends up to a few MHz.

4.2.3 *Wien bridge–based MMCC*

A Wien bridge–based MMCC [62, 67, 79, 82] is depicted in Fig. 4.16. As shown in the figure, the passive LC resonator part in the original MMCC was replaced with an active Wien bridge–based RC configuration. In addition to this main replacement, we implement the circuit as a fully inductorless form in the laboratory and present experimental results verifying the MMCC's both static and dynamic performance.

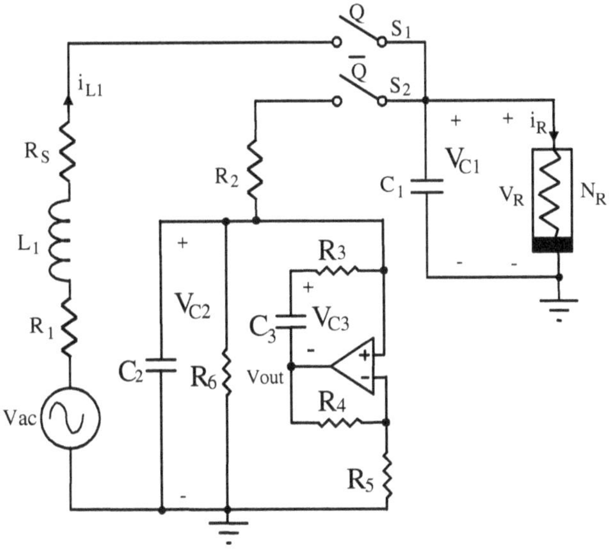

Fig. 4.16 Mixed-mode chaotic circuit with Wien bridge configuration.

Fig. 4.17 shows the experimental setup of an inductorless Wien bridge–based MMCC constructed with a CFOA-based Chua's diode, Wien bridge–based RC configuration and CFOA-based floating inductance simulator. In the experimental setup, Chua's diode is built by using two AD844 type CFOAs biased with ±9 V and four resistors, R_{N1} = R_{N2} = 22 kΩ, R_{N3} = 500 Ω, and R_{N4} = 2.2 kΩ. A floating inductance simulator block is configured by two AD844 type CFOAs biased with ±9 V; two resistors, $R_{F1} = R_{F2} = 1$ kΩ; and one capacitor, $C_F = 18$ nF, to simulate $L_1 = 18$ mH. Routine analysis for the CFOA-based floating grounded inductance simulator yields equivalent inductance as

$$L_{eq} = R_{F1} \cdot R_{F2} \cdot C_F \qquad (4.4)$$

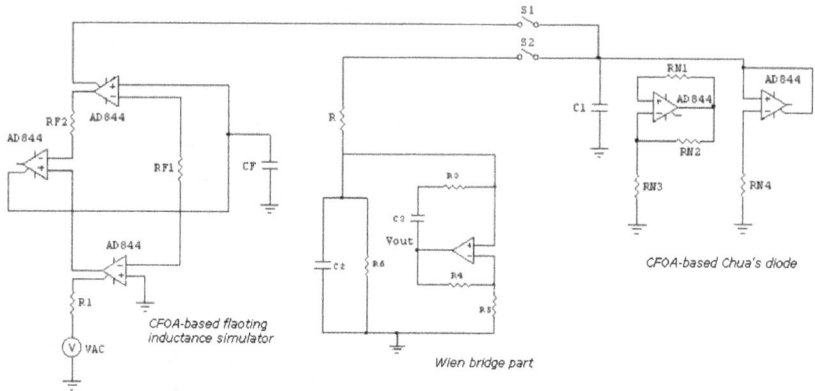

Fig. 4.17 Experimental setup of inductorless Wien bridge–based mixed-mode chaotic circuit.

A Wien bridge–based RC configuration was implemented using a VOA configured as a noninverting voltage-controlled voltage source with gain $A_v = [1+(R_4/R_5)]$. In this case, the circuit is described by the following set of three first-order autonomous differential equations

$$C_3 \frac{dV_{C3}}{dt} = -\frac{R_4}{R_3 R_5} V_{C2} - \frac{1}{R_3} V_{C3}$$

$$C_2 \frac{dV_{C2}}{dt} = \frac{1}{R_2} V_{C1} - \frac{1}{R_X} V_{C2} + \frac{1}{R_3} V_{C3} \qquad (4.5)$$

$$C_1 \frac{dV_{C1}}{dt} = \frac{1}{R_2} (V_{C2} - V_{C1}) - f(V_R)$$

where $f(V_R)$ has the same definition of Eq. (1.10) and

$$R_X = \frac{R_2 R_3 R_6 R_5}{R_2 (R_3 R_5 - R_6 R_4) + R_3 R_6 R_5} \tag{4.6}$$

We determined the circuit parameters of the Wien bridge–based RC configuration as $C_1 = 10$ nF, $C_2 = C_3 = 800$ nF, $R_3 = R_5 = R_6 = 100\ \Omega$, $R_4 = 207\ \Omega$ and $R_2 = 1.525$ kΩ, for which the circuit exhibits double-scroll chaotic attractor. Note that the theoretical gain required to start oscillations in a Wien oscillator with the equal-R ($R_3 = R_6 = R_0$) and the equal-C ($C_2 = C_3 = C_0$) design is $A_v = 3$. Hence, voltage gain, A_v was chosen slightly beyond this nominal value. In addition to the above arrangements for implementing a nonlinear resistor, floating inductance simulator and Wien bridge configuration, we fixed other nonautonomous circuit parameters as $R_1 = 1340\ \Omega$, the frequency $f = 8890$ Hz and the amplitude of ac source V_{ac} $A = 0.15$ V. A 4016 IC analog switching device, which contains four voltage-controlled analog switches, was used in our experiments as the electronic switching device.

4.2.3.1 *Experimental results*

In our experiments, we first verified the circuit's static performance. By applying a positive control signal to S1 and its inverse signal to S2 (nonautonomous mode), we measured the circuit's linear and nonlinear dynamics in time domain and X-Y projection using a digital storage oscilloscope interfaced with a computer.

In this operation mode, although the Wien bridge–based autonomous circuit part is not in the network, the Wien bridge circuit oscillates in sinusoidal form, *i.e.*, while nonautonomous chaotic oscillation is observed as voltage across the C1 capacitor in Fig. 4.17, periodic sinusoidal oscillation is observed in the Wien bridge circuit's output node simultaneously. The experimental observations including the MMCC circuit's nonautonomous chaotic behavior, the Wien bridge circuit part's output sinusoidal oscillation and double-scroll chaotic attractor in V_{C1}– $_{L1}$ projection are given in Fig. 4.18, respectively.

In the case of another switching process (S1-OFF, S2-ON), we have an autonomous Chua's circuit with a Wien bridge circuit part. For this operation mode, autonomous chaotic waveforms and double-scroll

Chua's attractor in V_{C1}–V_{C2} projection obtained from experiments are illustrated in Fig. 4.19.

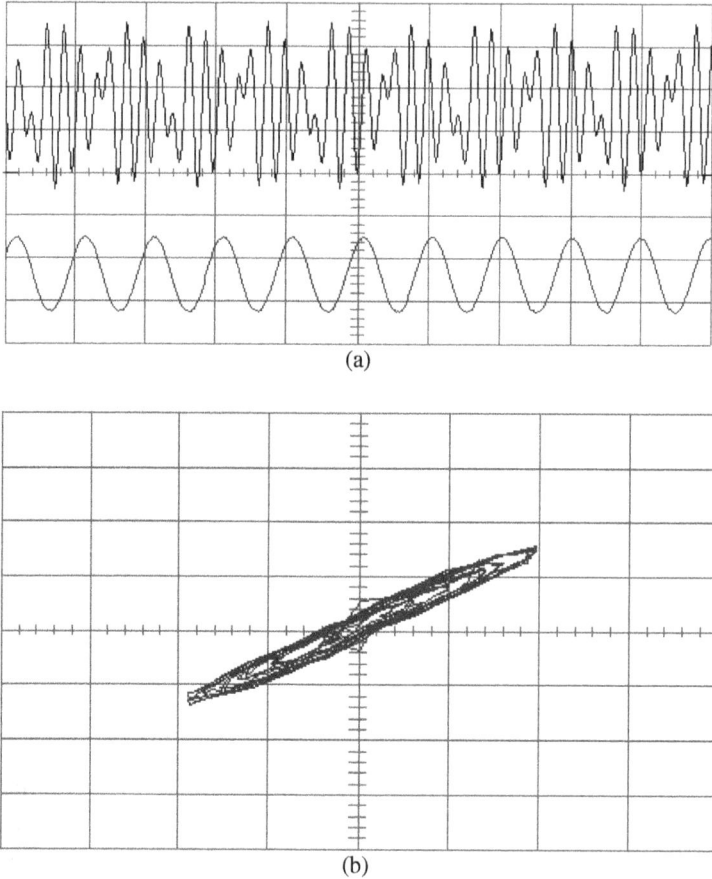

(a)

(b)

Fig. 4.18 (a) Chaotic and periodic dynamics of inductorless Wien bridge–based mixed-mode chaotic circuit in nonautonomous operation mode; the upper trace V_{C1} (2 V/div), the lower trace, output signal of Wien bridge circuit part, V out, (1 V/div), time/div: 500 μs/div, (b) chaotic attractor observed in the V_{C1}–i_{L1} plane, x-axes: 1 V, y-axes: 500 mV.

After experimentally confirming the static performance, in order to verify dynamic performance of the inductorless Wien bridge–based MMCC, we applied two complementary square waves Q and $\overline{Q}$, which have 5 ms impulse width and 10 ms period, in our experiments. The experimental results including the controlling signal and mixed-mode chaotic dynamic are shown in Fig. 4.20.

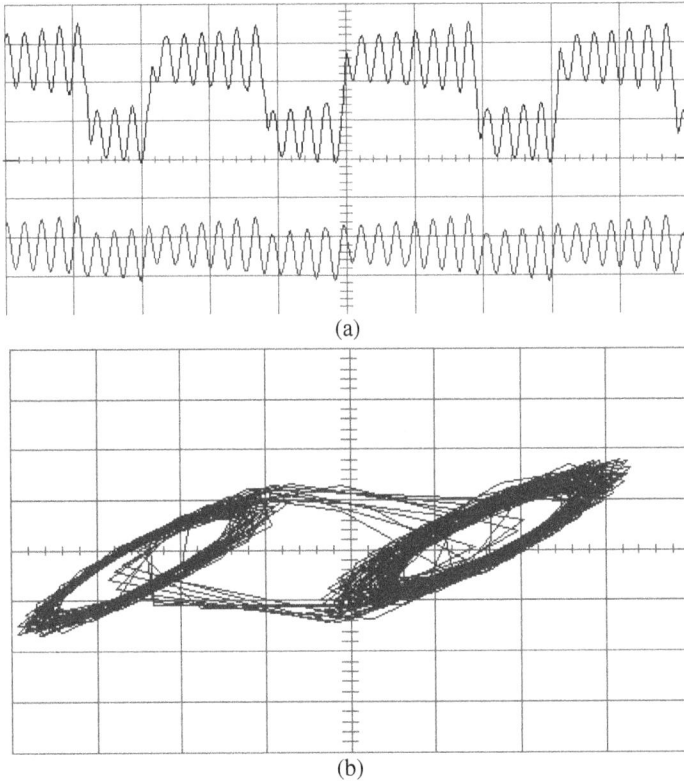

(a)

(b)

Fig. 4.19 (a) Chaotic circuit dynamics of inductorless Wien bridge–based mixed-mode chaotic circuit in autonomous operation mode; the upper trace V_{C1} (2 V/div), the lower trace V_{C2} (1 V/div), time/div: 2 ms/div, (b) chaotic attractor observed in the V_{C1}–V_{C2} plane, x-axes: 1 V, y-axes: 500 mV.

(a)

(b)

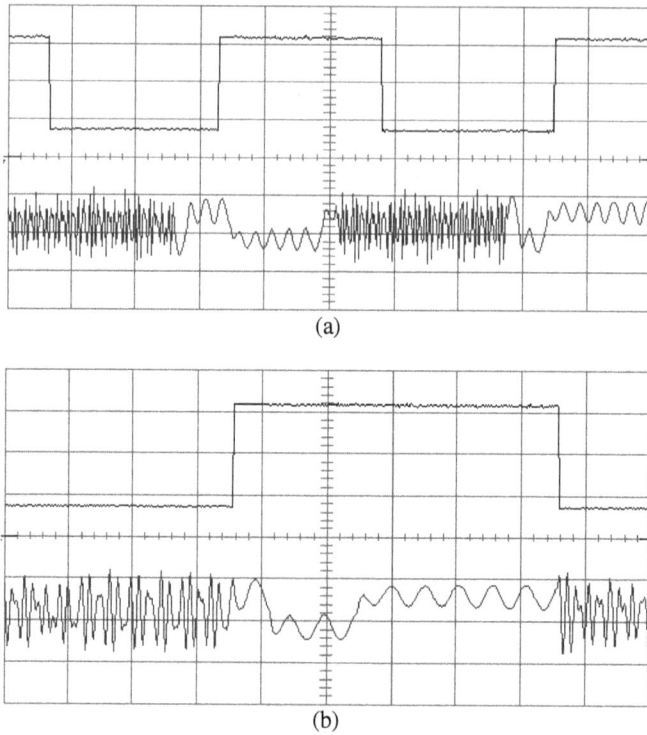

Fig. 4.20 Switching control signal and mixed-mode chaotic dynamics illustrations in Wien bridge–based MMCC circuit, (a) upper trace controlling signal (5 V/div), lower trace V_{C1} (5 V/div), time/div: 2 ms/div, (b) upper trace controlling signal (5 V/div), lower trace V_{C1} (5 V/div), time/div: 1 ms/div.

Chapter 5

Experimental Modifications of Chua's Circuits

In order to operate in higher dimensional form of autonomous and nonautonomous Chua's circuits keeping their original chaotic behaviors, we experimentally modified a VOA-based autonomous Chua's circuit and a nonautonomous MLC circuit by using a simple experimental method [65, 78]. In this chapter, after introducing this experimental method, we present PSPICE simulation and experimental results of modified high-dimensional autonomous and nonautonomous Chua's circuits.

5.1 Experimental Modifications of Autonomous and Nonautonomous Chua's Circuits

Our modifications are based on the circuit connections of a VOA-based Chua's diode [57]. The connection between a VOA-based Chua's diode and C_1 capacitor in an autonomous Chua's circuit and a nonautonomous MLC circuit as shown in Fig. 5.1 is the main modification point for higher-dimensional operation of the circuits. A modified Chua's circuit and MLC circuit are shown in Fig. 5.2(a) and (b), respectively. As shown in Fig. 5.2, the connections of resistors R_{N1} and R_{N4} from noninverting inputs of VOAs is broken, and an additional energy storage passive element (L_i or C_i) is inserted between the new common connection point (A-node) of R_{N1} and R_{N4} resistors, and common noninverting inputs (B-node) of VOAs. So, by inserting a passive element to Chua's diode structure, a path is provided between the new common connection point (A) of R_{N1} and R_{N4} resistors and their original connection point (B) with an impedance with respect to the inserted passive element.

(a)

(b)

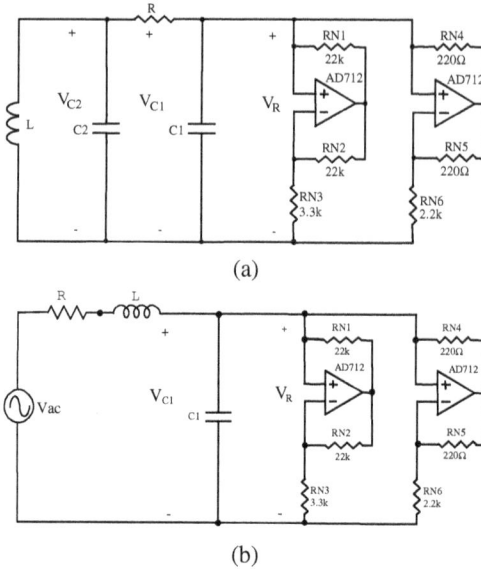

Fig. 5.1 (a) Chua's circuit using VOA-based Chua's diode, (b) MLC circuit using VOA-based Chua's diode.

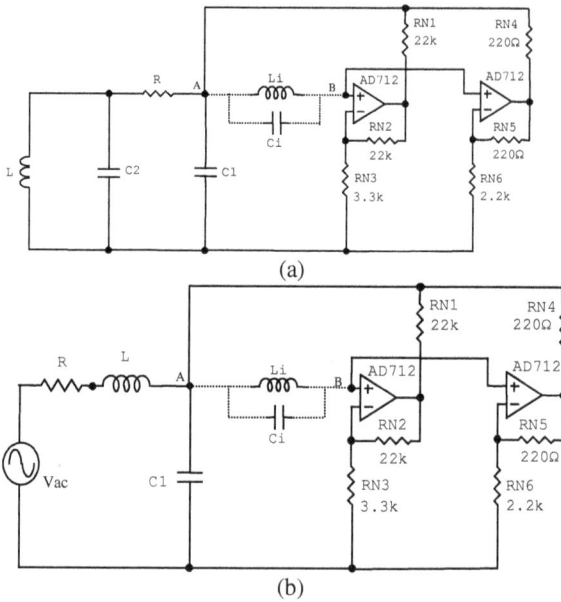

(a)

(b)

Fig. 5.2 Modification schemes of (a) Chua's circuit, (b) MLC circuit.

Although the structures of the circuits are modified with this interesting method, the circuits are still able to exhibit their original chaotic dynamics. With this modification, both the Chua's circuit and MLC circuit are turned into higher-dimensional chaotic form while keeping their original behaviors. After confirming this interesting modification method on both Chua's circuits with PSPICE simulations, we experimentally verified the modification scheme for higher-dimensional operation. In the next section, we give simulation and experimental results for modified Chua's circuits.

5.1.1 *Simulation results of modified Chua's circuits*

A guiding principle in our modification is that Chua's circuit structures must not be destroyed by inserting a passive element. Now, let's investigate the modified circuits in Fig. 5.2 with an inserted capacitive element, C_i. In this configuration, the inserted capacitor C_i provides a path between the defined nodes in Fig. 5.2 with an ac impedance according to $X_C = 1/2\pi f C$. If we choose a very small value of the capacitance, the circuits do not exhibit their chaotic behaviors, since the capacitive reactance will increase, leading to an open-circuit equivalent between the defined nodes. Simulation results of two circuits for $C_i = 1$ pF are shown in Fig. 5.3(a) and (b), respectively. By using nominal values of capacitance, which constitutes a path between the defined nodes, both circuits with this configuration exhibit chaotic oscillations as illustrated in Fig. 5.4(a) and (b) for Ci = 1 μF.

If we replace the inductor element L_i with capacitor C_i in the previous configuration, this inductor also provides a path between the defined connections with an ac impedance according to $X_L = 2\pi f L$. By choosing a very large value of inductance, the circuits do not exhibit their original chaotic behaviors since the large value of inductance causes an open-circuit equivalent between the defined connections. But if we use a suitable value of inductance, the circuits will exhibit their original chaotic behaviors. For the value of inserted inductor element $L_i = 15.10^4$ H and $L_i = 33$ mH, simulated circuit dynamics of two chaotic circuits are given in Fig. 5.5 and Fig. 5.6, respectively.

(a)

(b)

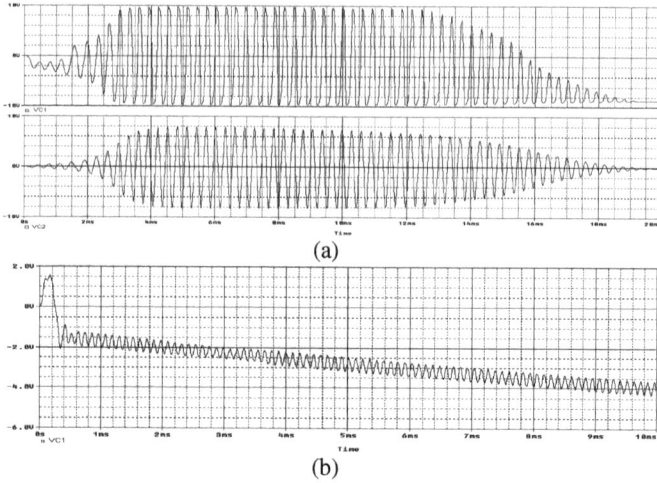

Fig. 5.3 PSPICE simulation results of modified Chua's circuit and MLC circuit with capacitive element C_i inserted between A- and B-nodes in Fig. 5.2 for a very small value of capacitance, $C_i = 1$ pF , (a) circuit dynamics of Chua's circuit, (b) circuit dynamic of MLC circuit.

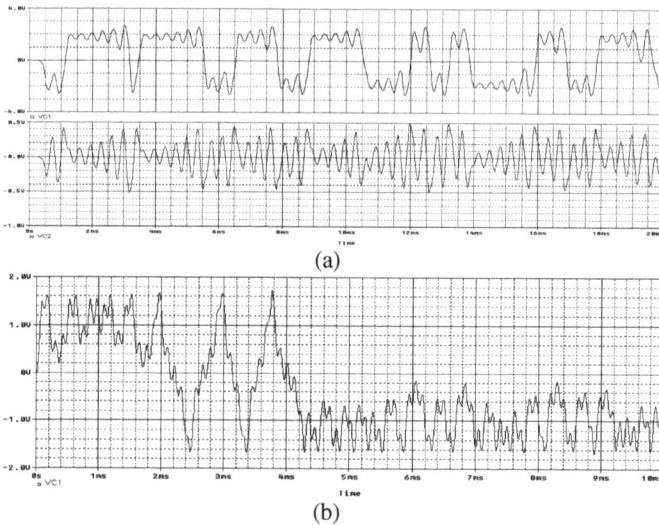

(a)

(b)

Fig. 5.4 PSPICE simulation results of modified Chua's circuit and MLC circuit with a capacitive element C_i inserted between A- and B-nodes in Fig. 5.2 for $C_i = 1$ μF, (a) chaotic dynamics of Chua's circuit, (b) chaotic dynamic of MLC circuit.

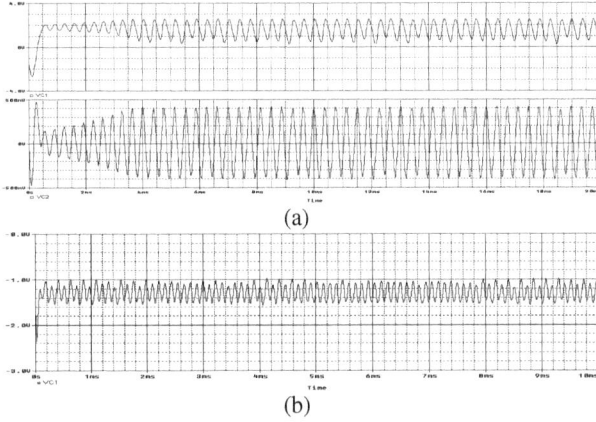

(a)

(b)

Fig. 5.5 PSPICE simulation results of modified Chua's circuit and MLC circuit with inductive element L_i inserted between A- and B-nodes in Fig. 5.2 for very large value of inductance, $L_i = 15 \times 10^4$ H, (a) circuit dynamics of Chua's circuit, (b) circuit dynamic of MLC circuit.

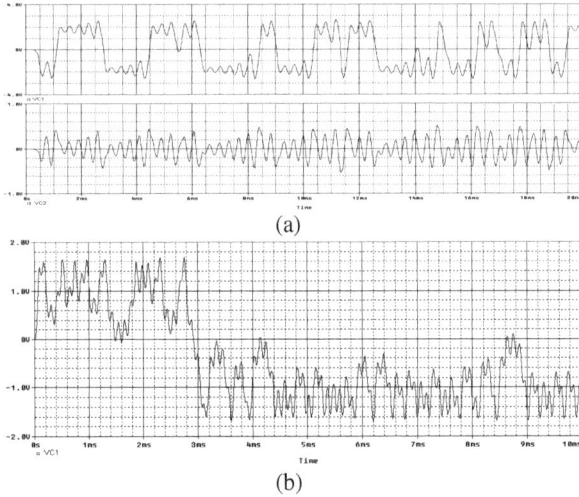

(a)

(b)

Fig. 5.6 PSPICE simulation results of modified Chua's circuit and MLC circuit with inductive element L_i inserted between A- and B-nodes in Fig. 5.2 for $L_i = 33$ mH, (a) chaotic dynamics of Chua's circuit, (b) chaotic dynamic of MLC circuit.

5.1.2 *Experimental results of modified Chua's circuits*

After confirming the design idea of modified Chua's circuits with two configurations with respect to two inserted energy storage passive elements, we have experimentally implemented the modified autonomous Chua's circuit and nonautonomous MLC circuit in the laboratory by constructing experimental setups. In these implementations, we used the classical parameter values C_1 = 10 nF, C_2 = 100 nF, L = 18 mH, R = 2 kΩ pot for Chua's circuit, and the parameters L = 18 mH, R = 2 kΩ pot, C = 10 nF, the frequency f = 8890 Hz, the amplitude A = 0.15 V of Vac for MLC circuit. We fixed the circuit parameter for Chua's diode in both circuits as R_{N1} = 22 kΩ, R_{N2} = 22 kΩ, R_{N3} = 3.3 kΩ, R_{N4} = 220 Ω, R_{N5} = 220 Ω, R_{N1} = 2.2 kΩ and two AD712 type VOAs biased with ±12 V.

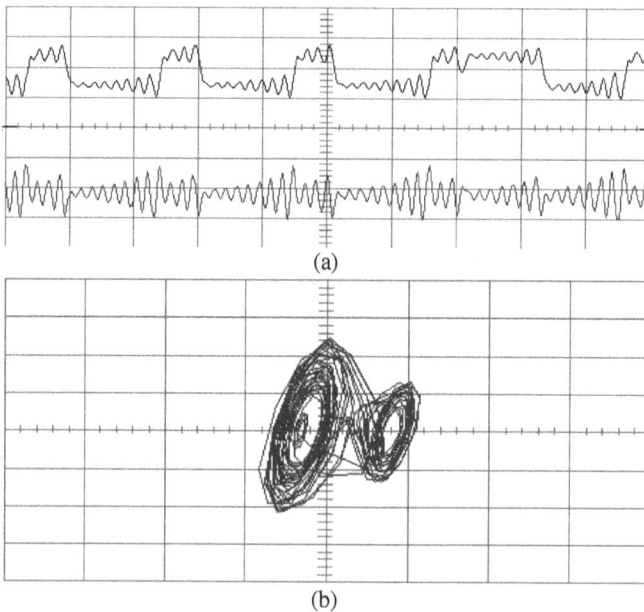

(a)

(b)

Fig. 5.7   Experimental results of modified Chua's circuit with capacitive element C_i inserted between A- and B-nodes in Fig. 5.2 for C_i = 1 μF, (a) chaotic circuit dynamics, the upper trace V_{C1} (5 V/div), the lower trace V_{C2} (1 V/div), time/div: 2 ms/div, (b) the chaotic attractor observed in the V_{C1}–V_{C2} plane, x-axes: 2 V, y-axes: 500 mV.

(a)

(b)

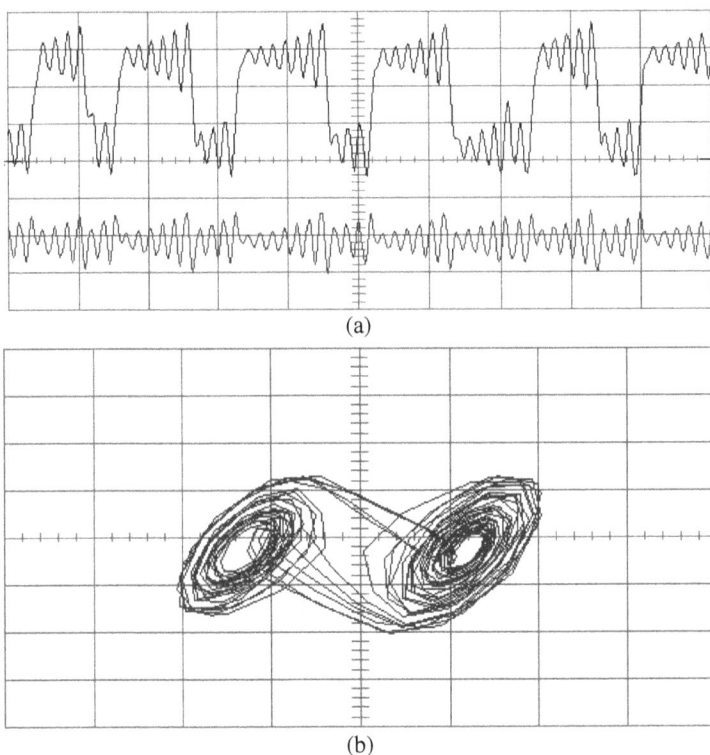

Fig. 5.8 Experimental results of modified Chua's circuit with inductive element L_i inserted between A- and B-nodes in Fig. 5.2 for $L_i = 33$-mH, (a) chaotic circuit dynamics, the upper trace V_{C1} (2-V/div), the lower trace V_{C2} (1-V/div), time/div: 2-ms/div, (b) the chaotic attractor observed in the V_{C1}–V_{C2} plane, x-axes: 2 V, y-axes: 500 mV.

The experimental results show good agreement with simulation results, and these experiments verified our modification schemes. From laboratory experiments, we determined that a Chua's circuit modified with inductor element L_i offers a more robust configuration than the other method. The experimental results of a modified Chua's circuit recorded in time domain and X-Y projections are illustrated in Fig. 5.7 and Fig. 5.8 for the modification scheme with $C_i = 1$ μF and modification scheme with $L_i = 33$ mH.

Fig. 5.9 shows the experimental results of an MLC circuit modified with the inserted inductor element L_i. These results include period-1 (Fig.

5.9a), period-2 (Fig. 5.9b), period-3 (Fig. 5.9c) and double-scroll chaotic behavior (Fig. 5.9d).

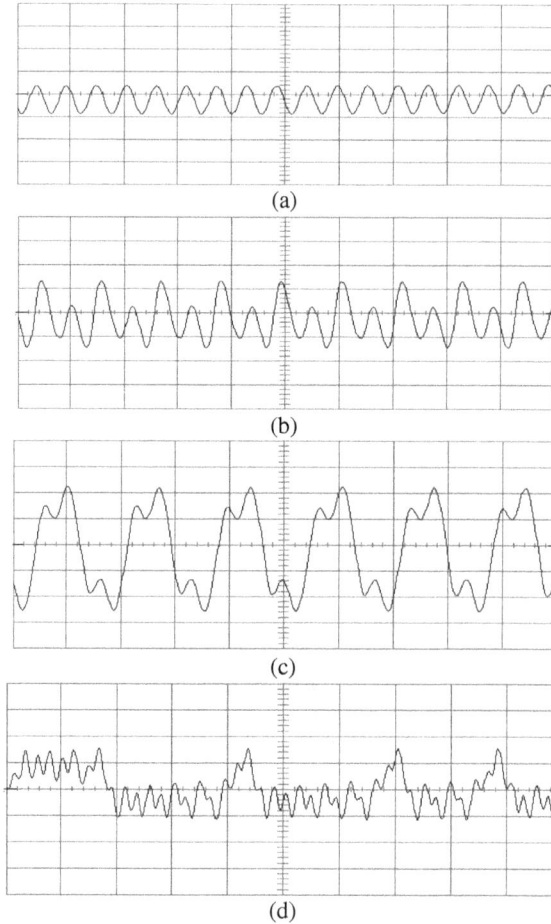

(a)

(b)

(c)

(d)

Fig. 5.9   Experimental results including period-doubling bifurcation sequence of modified MLC circuit with inductive element L_i inserted between A- and B-nodes in Fig. 5.2 for L_i = 33 mH, (a) A = 0.05 V; period-1 behavior of the trace V_{C1} (1 V/div), time/div: 200 µs/div (b) A = 0.1 V; period-2 behavior of the trace V_{C1} (1 V/div), time/div: 200 µs/div (c) A = 0.2 V; period-3, behavior of the trace V_{C1} (1 V/div), time/div: 200 µs/div (d) A = 0.15 V; double-band chaotic behavior of the trace V_{C1} (2 V/div), time/div: 500 µs/div.

The modified MLC circuit's double-scroll chaotic attractor was also obtained in our experiments. This two-dimensional attractor was measured in V_{C1}-$\dot{I}_L$ projection. We used a current-sensing resistor R_{sens} = 200 Ω for monitoring $\dot{I}_L$ current at oscilloscope screen in voltage form. These V_{C1} and $\dot{I}_L$ chaotic dynamics of the modified MLC circuit and double-scroll chaotic attractor are shown in Fig. 5.10.

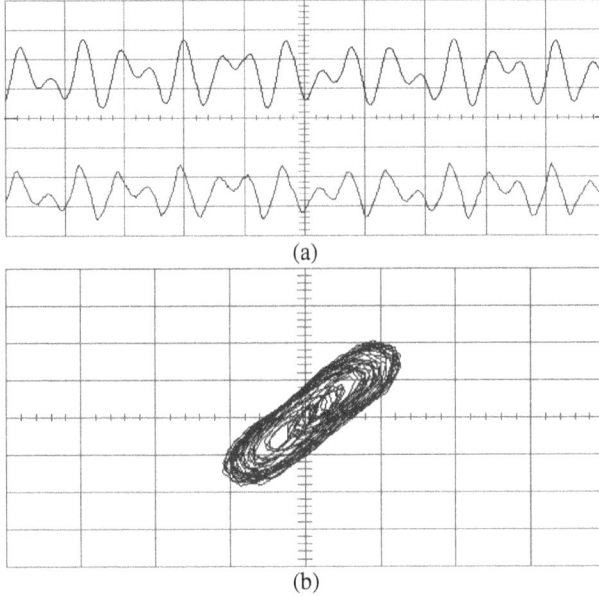

(a)

(b)

Fig. 5.10 Experimental results of modified MLC circuit with inductive element L_i inserted between A- and B-nodes in Fig. 5.2 for L_i = 33 mH, (a) chaotic circuit dynamics, the upper trace V_{C1} (2 V/div), the lower trace $\dot{I}_L$ in voltage form via a current-sensing resistor (200 mV/div), time/div: 2 ms/div, (b) the double-scroll chaotic attractor observed in the V_{C1}–$\dot{I}_L$ plane, x-axes: 2 V, y-axes: 100 mV.

5.2 A New Nonautonomous Version of VOA-Based Chua's Circuit

The proposed nonautonomous version of VOA-based Chua's circuit [78] is shown in Fig. 5.11. While the passive part is the same as that of an autonomous Chua's circuit, a sinusoidal excitation source V_{ac} is inserted in the nonlinear resistor structure.

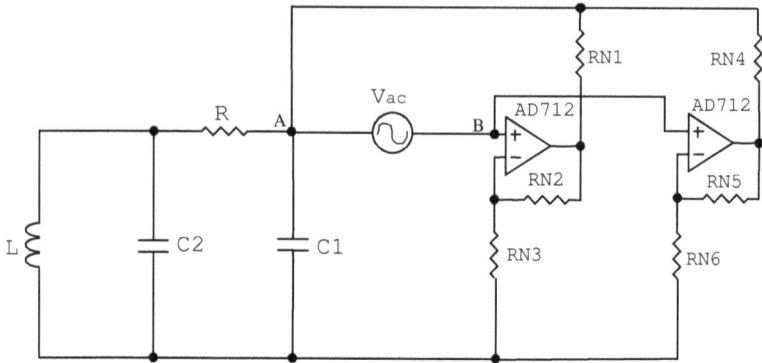

Fig. 5.11 A new nonautonomous version of Chua's circuit.

In this nonautonomous modification, the connections of resistors R_{N1} and R_{N4} from noninverting inputs of VOAs in the original Chua's circuit are broken, and an additional sinusoidal excitation source V_{ac} is inserted between the new common connection point (A-node) of R_{N1} and R_{N4} resistors and common noninverting inputs (B-node) of VOAs. By inserting an ac source to Chua's diode structure, a path is provided between the new common connection point (A-node) of the resistors and their original connection point (B-node). Since the characteristic features of this path are dependent on the frequency and amplitude of the inserted ac voltage source V_{ac}, this modification may be considered as a frequency/amplitude-controlled switch. Depending on frequency and amplitude parameters of the voltage source, the path forms the original connections of the resistors R_{N1} and R_{N4}, and the new nonautonomous version of Chua's circuit can successfully exhibit the original chaotic dynamics of an autonomous Chua's circuit.

The experimental setup for the new nonautonomous version of Chua's circuit in Fig. 5.11 was constructed by using the following circuit parameters: $C_1 = 10$ nF, $C_2 = 100$ nF, $R = 2$ kΩ pot, $R_{N1} = 22$ kΩ, $R_{N2} = 22$ kΩ, $R_{N3} = 3.3$ kΩ, $R_{N4} = 220$ Ω, $R_{N5} = 220$ Ω, $R_{N1} = 2.2$ kΩ, two AD712 type VOAs biased with ± 12 V. We determined these circuit parameters such that the circuit without ac voltage source (V_{ac}) exhibits double-band chaotic behavior. After constructing the modified circuit, we investigated the circuit behaviors by varying amplitude *(A)* of V_{ac} from zero to 500 mV and frequency *(f)* of V_{ac} in the range of 200 Hz–10

MHz. In addition to the experimental results, we also give PSPICE simulation results for comparison.

5.2.1 *Simulation results and experimental observations*

In our experiments, we first fixed the amplitude in the range of 100 mV–300 mV, and then we experimentally investigated the circuit for different frequency values of V_{ac}. For the defined amplitude level and frequency values less than 5 kHz, the circuit doesn't oscillate chaotically. By increasing the frequency value above 5 kHz, the circuit starts to behave chaotically. For f = 20 kHz and A = 100 mV, simulation and experimental results have been illustrated in Fig. 5.12 and Fig. 5.13, respectively.

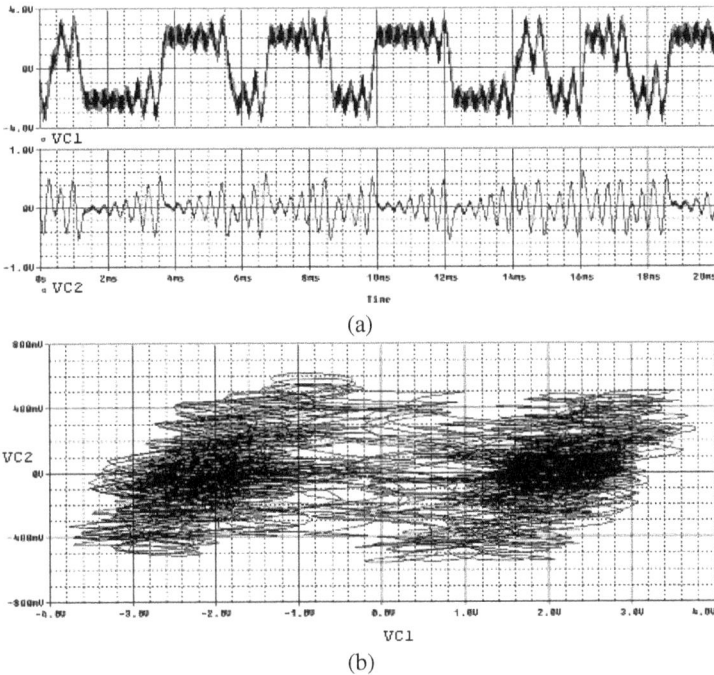

(a)

(b)

Fig. 5.12 PSPICE simulation results of nonautonomous circuit in Fig. 5.11 for f = 20 kHz, A = 100 mV of Vac.

As shown in the illustrations, the circuit dynamics have high-frequency components. As a fascinating result, in the case of further increasing (*f*) up to MHz levels, the circuit behaves with double-band chaotic dynamics but without high-frequency components. By keeping *A* = 100 mV, for 200 kHz, 2 MHz and 10 MHz, simulation and experimental results which include chaotic dynamics and double-scroll chaotic attractors have been given in Fig. 5.14, 5.15 and 5.16, respectively. After investigation of the proposed circuit for fixed amplitude level (100 mV) and 200 Hz–10 MHz frequency range, the effects of increasing amplitude levels greater than 300 mV were investigated.

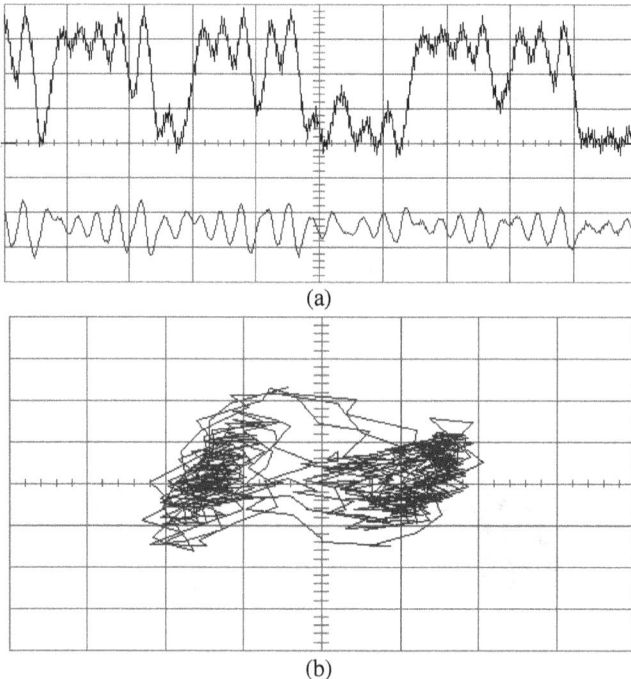

(a)

(b)

Fig. 5.13 Experimental results of nonautonomous circuit in Fig. 5.11 for *f* = 20 kHz, *A* = 100 mV of Vac, (a) illustrations of circuit dynamics, the upper trace V_{C1} (1 V/div), the lower trace V_{C2} (500 mV/div) time/div: 1 ms/div, (b) chaotic attractor projection in the V_{C1}–V_{C2} plane, x-axes: 1 V, y-axes: 200 mV.

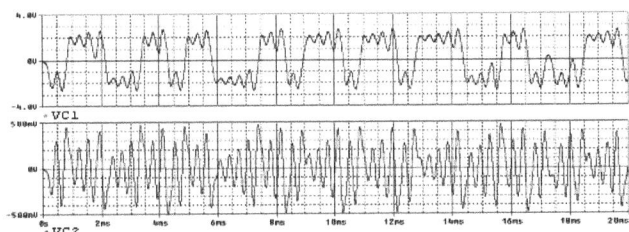

(a)

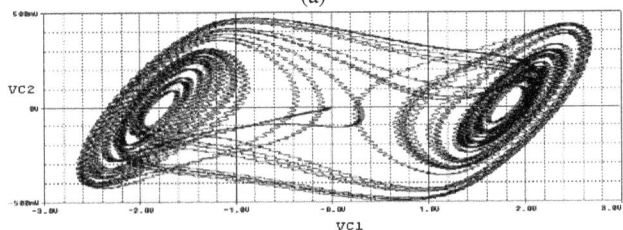

(b)

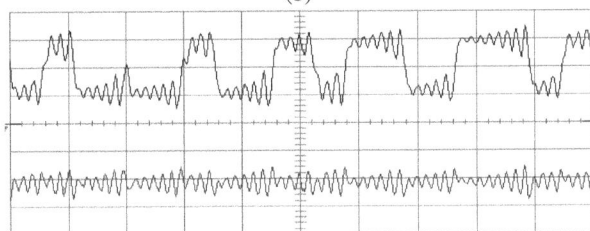

(c)

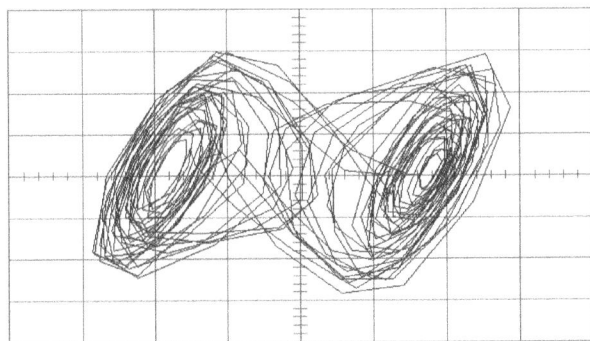

(d)

Fig. 5.14 PSPICE simulation and experimental results of nonautonomous circuit in Fig. 5.11, for $f = 200$ kHz, $A = 100$ mV of Vac, (a–b) simulation results include chaotic dynamics and chaotic attractor, (c) experimental observations of chaotic dynamics, the upper trace V_{C1} (2 V/div), the lower trace V_{C2} (1 V/div) time/div: 2 ms/div, (d) chaotic attractor projection in the V_{C1}–V_{C2} plane, x-axes: 1 V, y-axes: 200 mV.

(a)

(b)

(c)

(d)

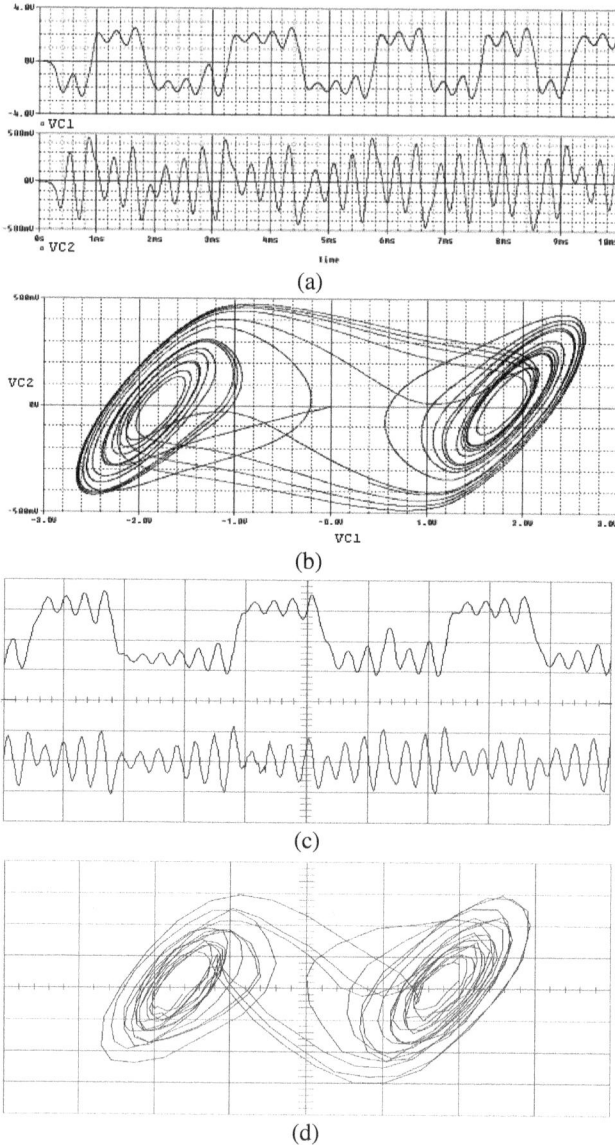

Fig. 5.15 PSPICE simulation and experimental results of nonautonomous circuit in Fig. 5.11, for f = 2 MHz, A = 100 mV of Vac, (a–b) simulation results include chaotic dynamics and chaotic attractor, (c) experimental observations of chaotic dynamics, the upper trace V_{C1} (2 V/div), the lower trace V_{C2} (500 mV) time/div: 1 ms/div, (d) chaotic attractor projection in the V_{C1}–V_{C2} plane, x-axes: 1 V, y-axes: 200 mV.

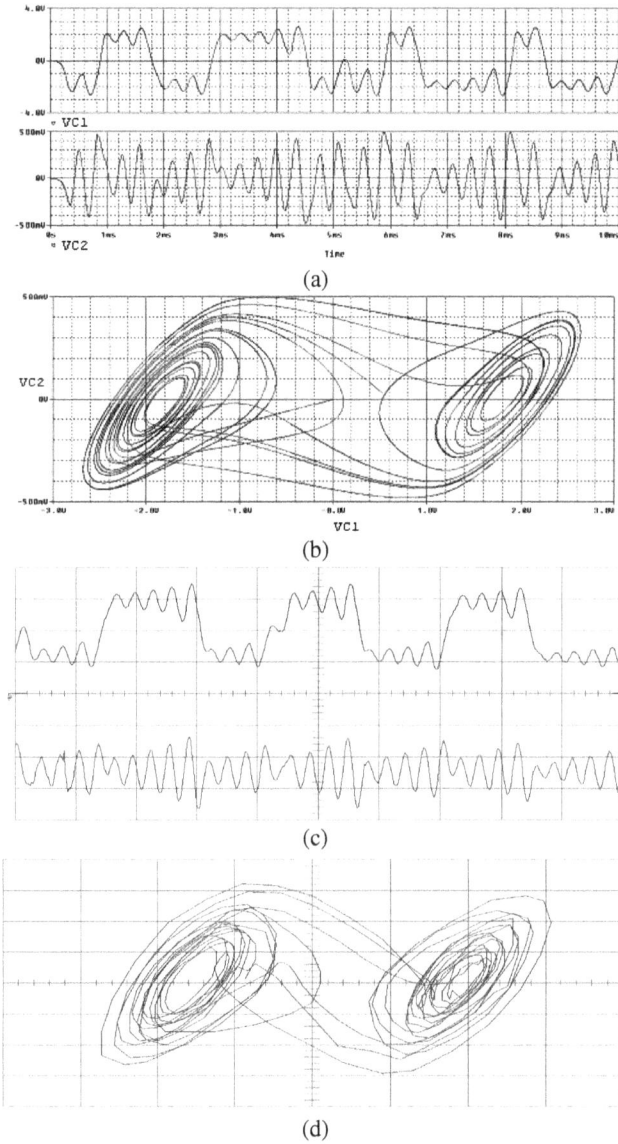

Fig. 5.16 PSPICE simulation and experimental results of nonautonomous circuit in Fig. 5.11, for $f = 10$ MHz, $A = 100$ mV of Vac, (a–b) simulation results include chaotic dynamics and chaotic attractor, (c) experimental observations of chaotic dynamics, the upper trace V_{C1} (2 V/div), the lower trace V_{C2} (500 mV) time/div: 1 ms/div, (d) chaotic attractor projection in the V_{C1}–V_{C2} plane, x-axes: 1 V, y-axes: 200 mV.

When we increase the amplitude of the ac source above 300 mV, the circuit's dynamics start to change significantly. The simulation and experimental results for one example of this operation mode are depicted in Fig. 5.17 for 2 MHz/500 mV parameter values.

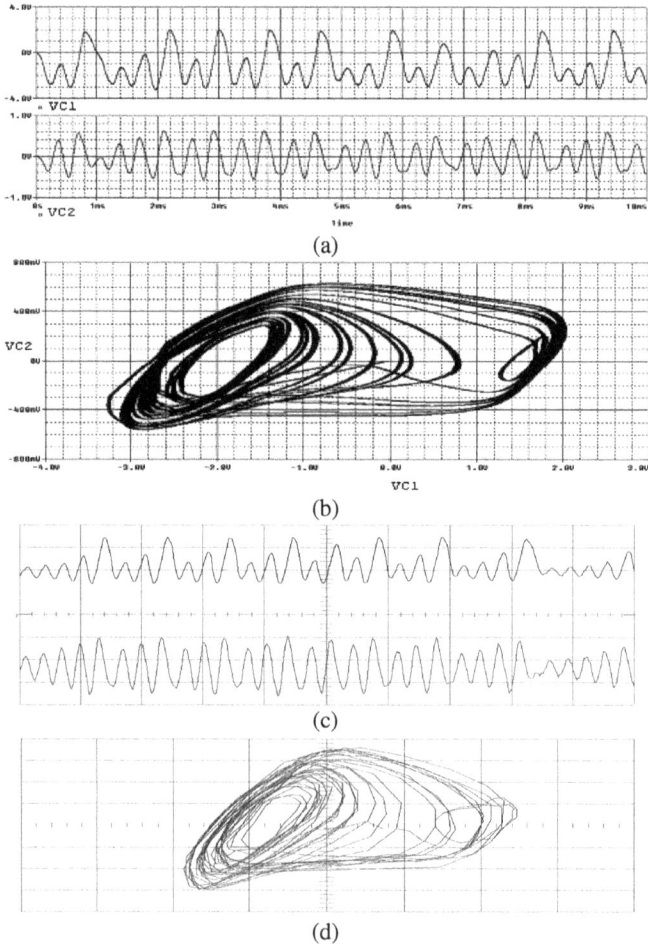

(a)

(b)

(c)

(d)

Fig. 5.17 PSPICE simulation and experimental results of nonautonomous circuit in Fig. 5.11, for f = 2 MHz, A = 500 mV of Vac, (a–b) simulation results include chaotic dynamics and chaotic attractor, (c) experimental illustrations of circuit dynamics; the upper trace V_{C1} (2 V/div), the lower trace V_{C2} (500 mV/div) time/div: 1 ms/div, (d) chaotic attractor projection in the V_{C1}–V_{C2} plane, x-axes: 1 V, y-axes: 200 mV.

5.3 Experimental Modification of MMCC

In this section, the modification method described in the former section was applied to the VOA-based MMCC. The proposed modified implementation is shown in Fig. 5.18. In this realization, the connections of resistors R_{N1} and R_{N4} from noninverting inputs of voltage op amps (VOAs) in the VOA-based MMCC are broken and an additional passive element (R_i, L_i or C_i) is inserted between the new common connection point (A-node) of resistors R_{N1} and R_{N4} and common noninverting inputs (B-node) of VOAs as shown in Fig. 5.18.

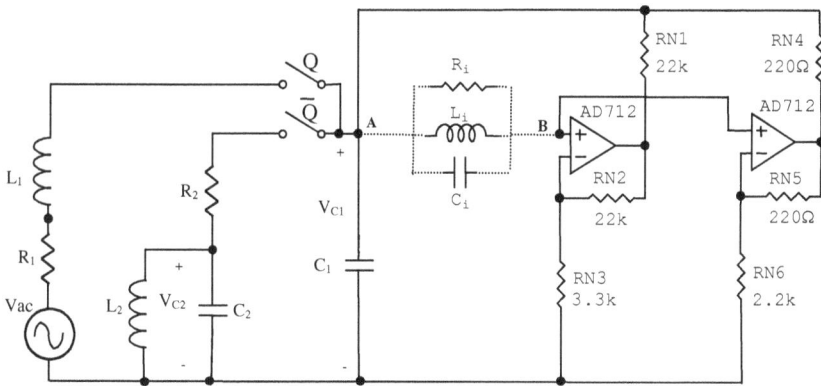

Fig. 5.18 New design scheme of MMCC.

This connection type provides a path between the new common connection point (A-node) of resistors R_{N1} and R_{N4} and their original connection point (B-node) with an impedance according to the inserted passive element. For example, if an inductor element is inserted between the defined connections (A- and B-nodes) in Fig. 5.18, this element provides a path between the defined connections with an ac impedance according to $X_L = 2\pi fL$. By choosing a very large value of inductance, the MMCC circuit doesn't exhibit its original chaotic dynamics, because the large value of inductance results in an open-circuit equivalent in the defined connections in Fig. 5.18. However, if we choose a suitable value of inductance, the circuit will exhibit the original mixed-mode chaotic circuit. We tested the proposed MMCC with three configurations according to the inserted passive element type by computer simulations and laboratory experiments. All configurations confirm our approach, and the laboratory experiments show good agreement with PSPICE simulations.

5.3.1 *Experimental results*

From laboratory experiments, we observed that the MMCC modified with inductor element is a more robust configuration than the others. We experimentally constructed this configuration by using the circuit parameters as $R_{N1} = R_{N2} = 22$ kΩ, $R_{N3} = 3.3$ kΩ, $R_{N4} = R_{N5} = 220$ Ω, $R_{N6} = 2.2$ kΩ, two AD712 type VOAs biased with ±12 V for Chua's diode, $R_1 = 1340$ Ω, $L_1 = L_2 = 18$ mH, $C_1 = 10$ nF, $C_2 = 100$ nF, $R_2 = 2$ kΩ pot, $L_i = 33$ mH, 4016IC switching device, the frequency f = 8890 Hz and the amplitude A = 0.15 V of Vac for other circuit autonomous and nonautonomous parts. We first verified the circuit's static performance using manual switching. By applying positive control voltage to S1 and negative control voltage to S2, we transformed the MMCC to the nonautonomous MLC circuit. In this operation mode, we measured the circuit's nonautonomous chaotic dynamics in time domain and x-y projection. By changing the control signals, *i.e.*, in case of switch positions S1-OFF and S2-ON, we have the autonomous Chua's circuit. The chaotic dynamics and double-scroll attractor behaviors of the modified MMCC for static mode are equivalent to the results given in Fig. 5.8 and Fig. 5.10.

In order to verify the dynamic performance of the new implementation, we applied two complementary square waves Q and $\overline{Q}$, which have 5 ms impulse width and 10 ms period, in our experiments. The experimental results including controlling signal, mixed-mode chaotic dynamic and chaotic attractor projection in V_{C2}–V_{C1} plane are illustrated in Fig. 5.19. In the experiments for dynamic operation mode of the modified MMCC, we measured the different switching projections. These projections are shown in Fig. 5.20, Fig. 5.21 and Fig. 5.22.

(a)

(b)

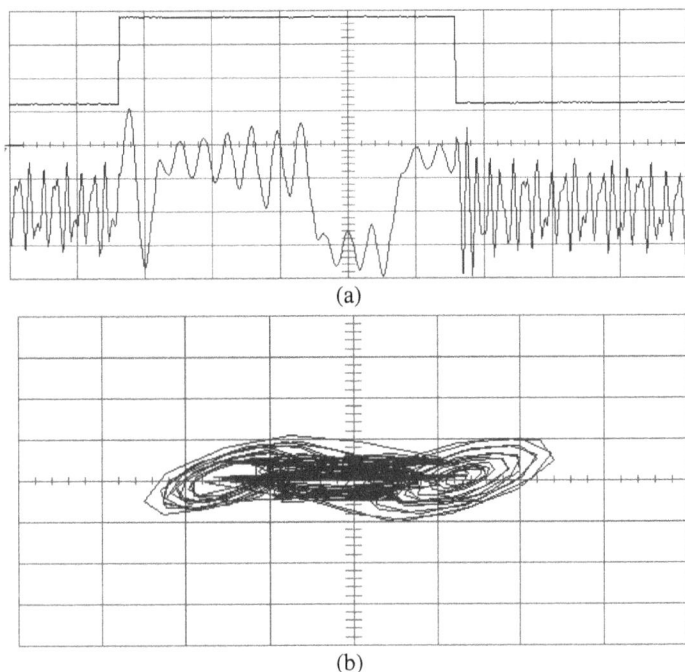

Fig. 5.19 Experimental results of new MMCC circuit in Fig. 5.18 for dynamic operation mode, (a) upper trace is controlling signal $\overline{Q}$ (5 V/div), lower trace is V_{C1}, (2 V/div), time/div: 1 ms/div, (b) the chaotic attractor observed in the V_{C1}–V_{C2} plane, x-axes: 2 V, y-axes: 1 V.

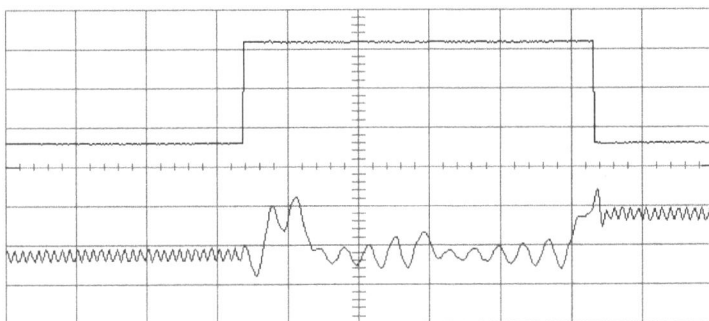

Fig. 5.20 The mixed-mode chaotic circuit combined with double-scroll autonomous dynamic and period-1 nonautonomous dynamic; the upper trace controlling signal (5V/div), the lower trace V_{C1} (5V/div), time/div: 1ms/div.

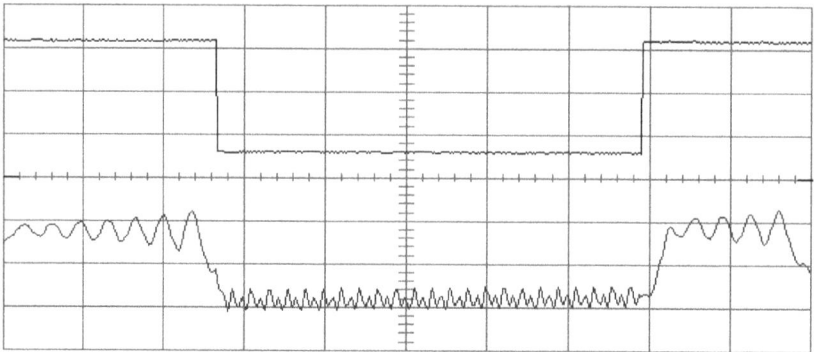

Fig. 5.21 The mixed-mode chaotic circuit combined with double-scroll autonomous dynamic and period-2 nonautonomous dynamic; the upper trace controlling signal (5V/div), the lower trace V_{C1} (5V/div), time/div: 1ms/div.

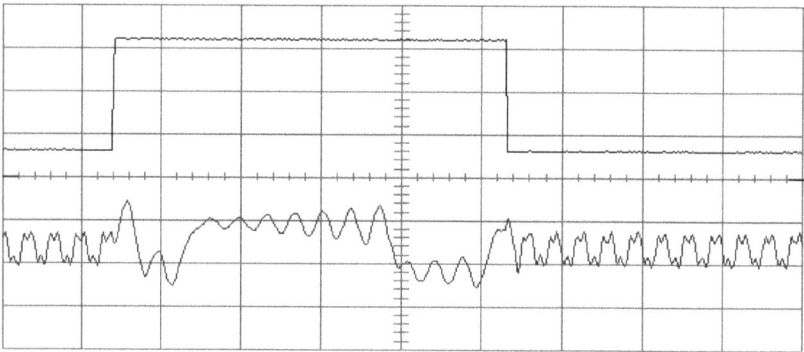

Fig. 5.22 The mixed-mode chaotic circuit combined with double-scroll autonomous dynamic and period-3 nonautonomous dynamic; the upper trace controlling signal (5V/div), the lower trace V_{C1} (5V/div), time/div: 1ms/div.

Chapter 6

Some Interesting Synchronization Applications of Chua's Circuits

In the literature, a lot of research has been devoted to synchronization issues of Chua's circuits [see for example 20, 26–28, 31, 37–38, 40, 43, 46, 48–50, 88, 100, 108, 110, 115, 125–126, 129, 134, 144, 147–148, 151, and references therein]. In this chapter, we also discuss some interesting synchronization applications of Chua's circuits. These applications include analog and digital communication system applications using mixed-mode chaotic circuits and continuous and impulsive synchronization applications of state-controlled cellular neural network (SC-CNN)-based circuit, which is a different version of the Chua's circuit.

6.1 An Analog Communication System Using MMCC

A communication system via the mixed-mode chaotic circuit [69] is shown in Fig. 6.1. This system is based on the chaotic modulation and demodulation technique proposed by Itoh *et al.* [54]. Depending on the node-1 voltage $V_R(t)$ that controls the switching state, the circuit equations for the transmitting system are given by

$$C_1 \frac{dV_R}{dt} = i_{L1} - f(V_R) + \frac{V_i(t) - V_R}{R_i}$$
$$L_1 \frac{di_{L1}}{dt} = -i_{L1}(R_1 + R_{S1}) - V_R + A \sin(wt)$$
$$, V_R \rangle 0 \quad (6.1)$$

and

$$L_2 \frac{di_{L2}}{dt} = -V_{C2} - i_{L2} \cdot R_{S2}$$

$$C_2 \frac{dV_{C2}}{dt} = i_{L2} - \frac{1}{R_2}\left(V_{C2} - V_R\right) \qquad\qquad , V_R \langle 0 \quad (6.2)$$

$$C_1 \frac{dV_R}{dt} = \frac{1}{R_2}\left(V_{C2} - V_R\right) - f\left(V_R\right) + \frac{V_i(t) - V_R}{R_i}$$

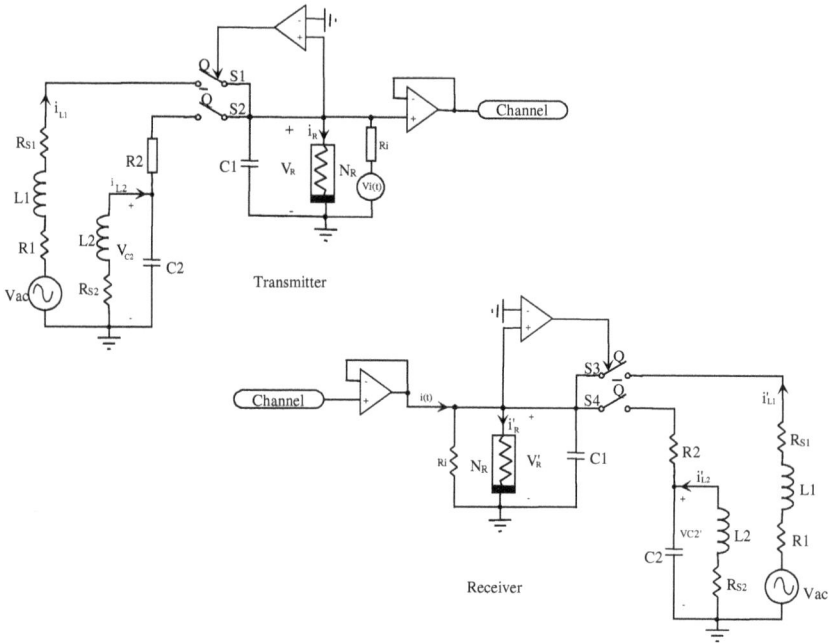

Fig. 6.1   Communication system based on chaotic modulation and demodulation technique, which utilizes mixed-mode chaotic circuits.

We use the voltage source $V_i(t)$ as the information signal, and $V_R(t)$ as the transmitted signal. The common dynamic of two switching case (i.e., $V_R \rangle 0$ and $, V_R \langle 0$) is rewritten as follows:

$$C_1 \frac{dV_R}{dt} = \frac{1}{2}\left[\begin{array}{l} (1+\mathrm{sgn}(V_R))\cdot(i_{L1} - f(V_R)) + \\ (1-\mathrm{sgn}(V_R))\cdot\left(\dfrac{V_{C2} - V_R}{R_2} - f(V_R)\right) \end{array} \right] + \frac{V_i(t) - V_R}{R_i} \quad (6.3)$$

The circuit equations for the receiver are given by

$$C_1 \frac{dV_R'}{dt} = i_{L1}' - f(V_R') - \frac{V_R'}{R_i}$$

$$L_1 \frac{di_{L1}'}{dt} = -i_{L1}'(R_1 + R_{S1}) - V_R' + A\sin(wt) \qquad , V_R' \rangle 0 \quad (6.4)$$

and

$$L_2 \frac{di_{L2}'}{dt} = -V_{C2}' - i_{L2}' \cdot R_{S2}$$

$$C_2 \frac{dV_{C2}'}{dt} = i_{L2}' - \frac{1}{R_2}(V_{C2}' - V_R') \qquad , V_R' \langle 0 \quad (6.5)$$

$$C_1 \frac{dV_R'}{dt} = \frac{1}{R_2}(V_{C2}' - V_R') - f(V_R') - \frac{V_R'}{R_i}$$

where $V_R(t) = V_R'(t)$ because of the voltage buffer. We can also rewrite common dynamic $C_1 \dfrac{dV_R'}{dt}$ for receiving system as:

$$C_1 \frac{dV_R'}{dt} = \frac{1}{2}\left[\begin{array}{c}(1+\text{sgn}(V_R'))\cdot(i_{L1} - f(V_R')) + \\ (1-\text{sgn}(V_R'))\cdot\left(\dfrac{V_{C2}'-V_R'}{R_2} - f(V_R')\right)\end{array}\right] - \frac{V_R'}{R_i} \quad (6.6)$$

In this step, we show that how the information signal can be recovered. From the equation (6.3), we obtain

$$V_i(t) = R_i\left[C_1\frac{dV_R}{dt} - \frac{1}{2}\left[\begin{array}{c}(1+\text{sgn}(V_R))\cdot(i_{L1} - f(V_R)) + \\ (1-\text{sgn}(V_R))\cdot\left(\dfrac{V_{C2}-V_R}{R_2} - f(V_R)\right)\end{array}\right] + \frac{V_R}{R_i}\right] \quad (6.7)$$

Similarly, the current $i(t)$ in Fig. 6.1 is given by

$$i(t) = \left[C_1\frac{dV_R'}{dt} - \frac{1}{2}\left[\begin{array}{c}(1+\text{sgn}(V_R'))\cdot(i_{L1}' - f(V_R')) + \\ (1-\text{sgn}(V_R'))\cdot\left(\dfrac{V_{C2}'-V_R'}{R_2} - f(V_R')\right)\end{array}\right] + \frac{V_R'}{R_i}\right] \quad (6.8)$$

Defining the difference $p(t) = v_{C2}(t) - v'_{C2}(t)$, $q(t) = i_{L2}(t) - i'_{L2}(t)$ and $r(t) = i_{L1}(t) - i'_{L1}(t)$, we get

$$C_2 \frac{dp}{dt} = -\frac{p}{R_2} + q$$

$$L_2 \frac{dq}{dt} = -p - R_{S1} \cdot q \qquad (6.9)$$

$$L_1 \frac{dr}{dt} = -(R_1 + R_{S1}) \cdot r$$

Since the origin of Eq. (6.9) is global-asymptotically stable, we can state that $|p| = |v_{C2} - v'_{C2}| \to 0$, $|q| = |i_{L2} - i'_{L2}| \to 0$ and $|r| = |i_{L1} - i'_{L1}| \to 0$ as $t \to \infty$. Thus, $i(t)$ is given by

$$i(t) = \left[C_1 \frac{dV'_R}{dt} - \frac{1}{2} \begin{bmatrix} (1 + \mathrm{sgn}(V'_R)) \cdot (i'_{L1} - f(V'_R)) + \\ (1 - \mathrm{sgn}(V'_R)) \cdot \left(\dfrac{V'_{C2} - V'_R}{R_2} - f(V'_R) \right) \end{bmatrix} + \frac{V'_R}{R_i} \right] \to \frac{V_i(t)}{R_i} \quad (6.10)$$

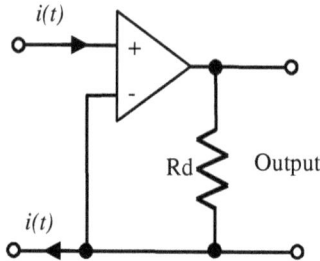

Fig. 6.2 Current detector.

Eq. (6.10) implies that the current $i(t)$ varies in proportion to the informational signal $V_i(t)$. Thus, if we use the current detector in Fig. 6.2, we can recover the information signal. In simulation experiments, we used $R_D = 2$ kΩ and $R_i = 33.9$ kΩ.

6.1.1 *Simulation results*

We have examined the performance of the secure communication system in Fig. 6.1 by computer simulation experiments. We used sine and triangle waves as the information signal. When the transmitter was modulated with an input signal $V_i(t)$ and the receiver had the same autonomous, nonautonomous parameters and switching method of the transmitter, chaotic synchronization between two systems was observed. The information signal was then recovered successfully at the receiver stage. Evidence of the receiver synchronization is shown in Fig. 6.3. The transmitted, input and output signals are shown in Fig. 6.4 and 6.5 for sine wave and triangle wave inputs, respectively. These figures demonstrate the effectiveness of the chaotic communication system using the mixed-mode chaotic circuit for different inputs. To eliminate high-frequency components observed in switching transitions, we used a low-pass filter in receiver stage.

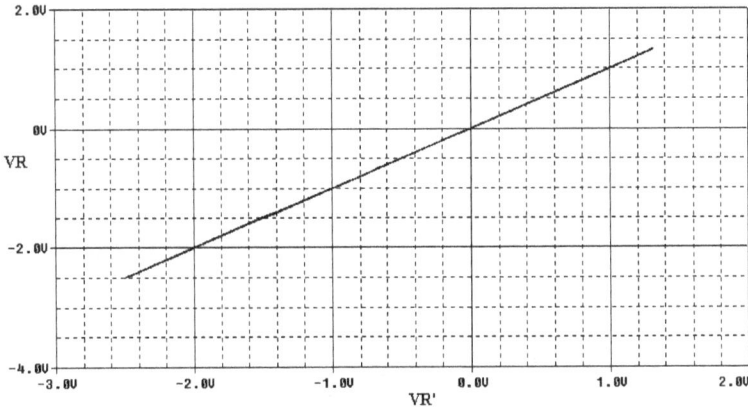

Fig. 6.3 Synchronization graph in communication system in Fig. 6.1.

6.2 Chaotic Switching System Using MMCC

A chaotic switching system using an MMCC [70] is shown in Fig. 6.6. As binary data stream, either V(Q) signal which controls the switch S1 or

$V(\overline{Q})$ signal which controls the switch S2 can be chosen. The binary data stream modulates chaotic carrier $V_{C1}(t)$.

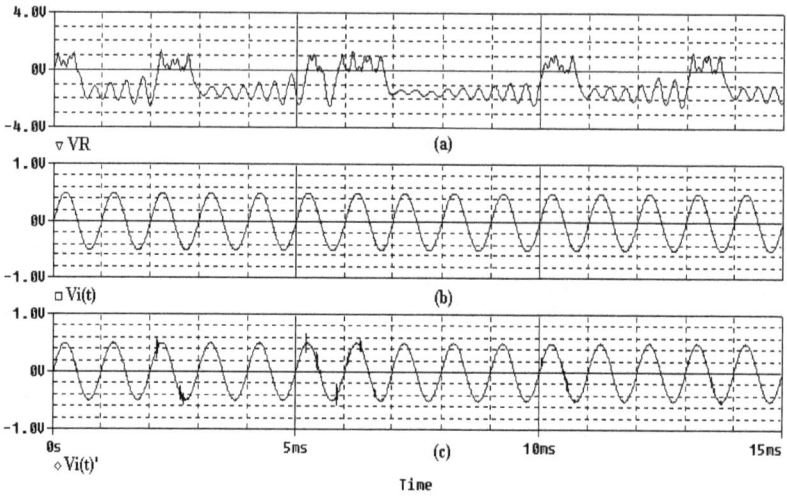

Fig. 6.4 (a) Transmitted mixed-mode chaotic signal with sine-wave input, (b) input signal of frequency 1 KHz and amplitude 0.5 V, (c) recovered signal in the amplified form.

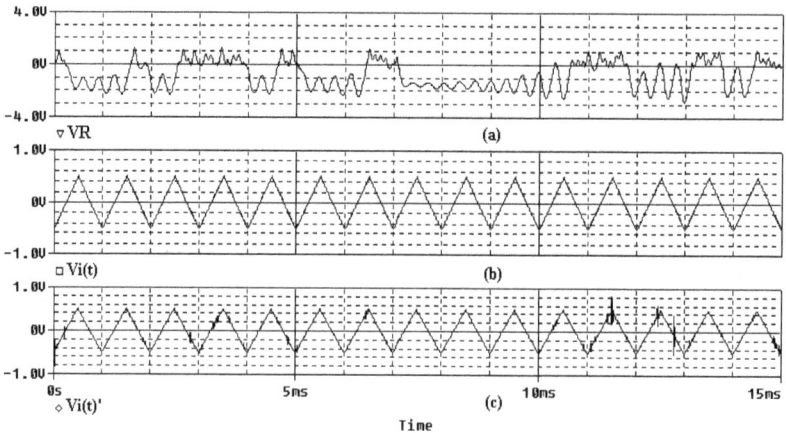

Fig. 6.5 (a) Transmitted mixed-mode chaotic signal with triangle-wave input, (b) input signal of frequency 1 KHz and amplitude 0.5 V, (c) recovered signal in the amplified form.

If an input bit +1 has to be transmitted, the switch S1 is closed (S1-ON and S2-OFF) for a time interval T. If the next bit −1 has to be transmitted, the switch S1 is kept open (S1-OFF and S2-ON).When the states of switches are S1-ON and S2-OFF (*i.e.*, input bit +1 is transmitted) in Fig. 6.6, we have the standard nonautonomous MLC chaotic circuit exhibiting a double-scroll chaotic attractor. In this case, the transmitter is represented by the following set of two first-order nonautonomous differential equations

$$C_1 \frac{dV_{C1}}{dt} = i_{L1} - f(V_{C1})$$

$$L_1 \frac{di_{L1}}{dt} = -i_{L1}(R_1 + R_{S1}) - V_{C1} + A\sin(wt) \qquad (6.11)$$

where *(A)* is the amplitude and *(w)* is the angular frequency of the external periodic force V_{ac} in Fig. 6.6.

When the states of switches are S1-OFF and S2-ON (*i.e.*, input bit −1 is transmitted), we have the standard autonomous Chua's circuit exhibiting a double-scroll Chua's chaotic attractor. In this case, the transmitter is described by the following set of first-order autonomous differential equations:

$$L_2 \frac{di_{L2}}{dt} = -V_{C2} - i_{L2} \cdot R_{S2}$$

$$C_2 \frac{dV_{C2}}{dt} = i_{L2} - \frac{1}{R_2}(V_{C2} - V_{C1}) \qquad (6.12)$$

As shown in Fig. 6.6, the receiver module is made of two subsystems which have the same circuit structures of the transmitter's autonomous and nonautonomous circuit parts. The first subsystem of the receiver is governed by the following equations:

$$C_1 \frac{dV_{C12}}{dt} = i'_{L1} - f(V_{C12}) + \frac{1}{R_C}(V_{C1} - V_{C12})$$

$$L_1 \frac{di'_{L1}}{dt} = -i'_{L1}(R_1 + R_{S1}) - V_{C12} + A\sin(wt) \qquad (6.13)$$

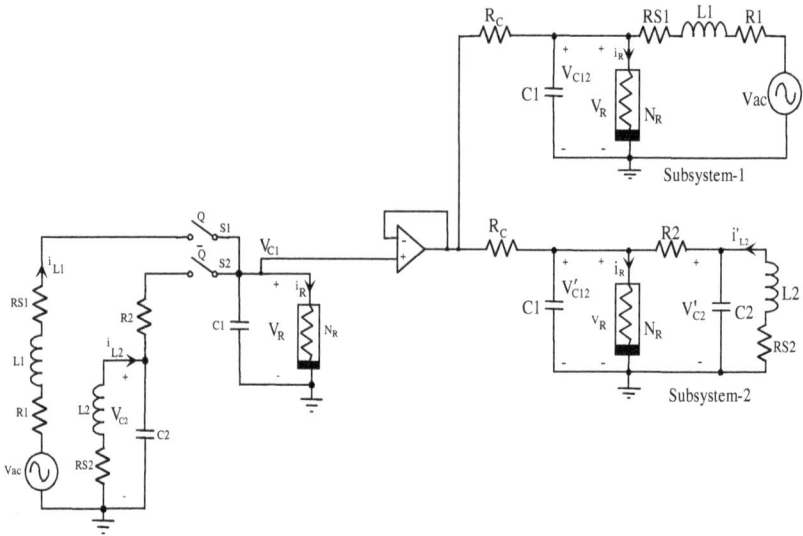

Fig. 6.6 A chaotic switching system using MMCC.

The second subsystem of the receiver is governed by the following equations:

$$L_2 \frac{di'_{L2}}{dt} = -V'_{C2} - i'_{L2} \cdot R_{S2}$$

$$C_2 \frac{dV'_{C2}}{dt} = i'_{L2} - \frac{1}{R_2}\left(V'_{C2} - V'_{C12}\right) \qquad (6.14)$$

$$C_1 \frac{dV'_{C12}}{dt} = \frac{1}{R_2}\left(V'_{C2} - V'_{C12}\right) - f\left(V'_{C12}\right) + \frac{1}{R_C}\left(V_{C1} - V'_{C12}\right)$$

Here, we used the synchronization method to achieve chaos synchronization between two identical nonlinear systems due to the effect of a one-way coupling element without any additional stable subsystem. In the receiver, $V_{C12}(t)$ converges to $V_{C1}(t)$ during the transmission of +1 bit, while $V'_{C12}(t)$ converges to $V_{C1}(t)$ during the transmission of −1 bit.

6.2.1 *Simulation results*

In our simulation experiments, the component values for the linear parts of the transmitter and receiver were chosen as follows: $L_1 = L_2 = 18$ mH, $C_1 = 10$ nF, $C_2 = 100$ nF, $R_1 = 1340$ Ω, $R_2 = 1700$ Ω, $R_{S1} = R_{S2} = 12.5$ Ω, $R_C = 1000$ Ω. The amplitude and frequency of the external periodic force V_{ac} are A = 0.15 V and f = 8890 Hz. We first show and verify the static performance of the chaotic switching system via the MMCC. When the chaotic signal corresponding to a +1 is transmitted, the receiver's first subsystem synchronizes with the received signal, while the second does not. In the case of the transmission of +1 bit, phase portraits for the receiver's two subsystems are shown in Fig. 6.7(a) and (b). When the chaotic signal corresponding to a −1 bit is transmitted, the reverse behavior is observed, *i.e.*, the second subsystem synchronizes with the received signal, while the first does not. In the case of the transmission of a −1 bit, phase portraits for the receiver's subsystems are shown in Fig. 6.7(c) and (d). From the phase portraits, we can see that the receiver can distinguish the different states correctly.

In order to evaluate the dynamic performance of chaotic switching system via the MMCC, we applied a square wave with 200 Hz ±1 sequence to the gate of the analog switch S1 and the complementary signal of the square wave with the same frequency to the gate of the analog switch S2 in Fig. 6.6. At the receiver, two synchronization errors, $e(t)$ and $e'(t)$, are produced between the received drive signal $V_{C1}(t)$ and the receiver's regenerated drive signals $V_{C12}(t)$ and $V'_{C12}(t)$. Using the synchronization error, the binary information signal can be recovered. For this purpose, we constructed a detection circuit shown as a block diagram in Fig. 6.8.

Simulation results are shown in Fig. 6.9. The figure shows binary input signal V(Q), transmitted signal $V_{C1}(t)$, error signal $e(t)$ between $V_{C1}(t)$ and $V_{C12}(t)$, error signal $e'(t)$ between $V_{C1}(t)$ and $V'_{C12}(t)$, and recovered binary signal V'(Q), respectively. Fig. 6.9 illustrates that the binary information signal can be reliably recovered by using the detection circuitry in Fig. 6.8.

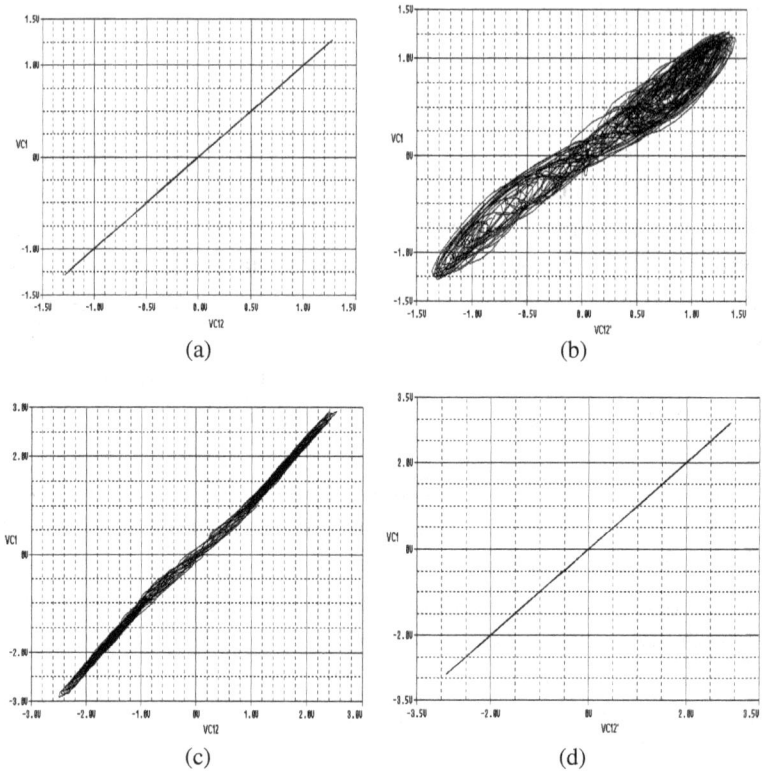

(a)

(b)

(c)

(d)

Fig. 6.7 Phase portraits for transmission +1 bit, (a) $V_{C12}(t)$ synchronizes with $V_{C1}(t)$, (b) $V'_{C12}(t)$ doesn't synchronize with $V_{C1}(t)$. Phase portraits for transmission −1 bit, (c) $V_{C12}(t)$ doesn't synchronize with $V_{C1}(t)$, (d) $V'_{C12}(t)$ synchronizes with $V_{C1}(t)$.

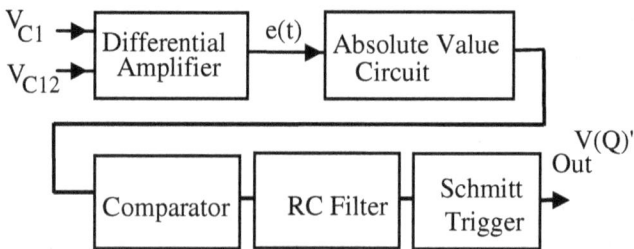

Fig. 6.8 Block diagram of detection circuit.

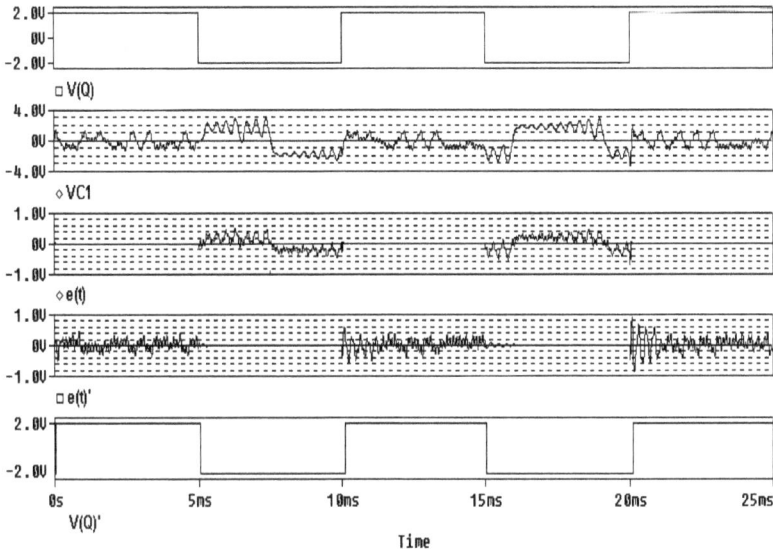

Fig. 6.9 Simulation results of the chaotic switching system in Fig. 6.6 for dynamic operation.

6.3 Chaos Synchronization in SC-CNN-Based Circuit and an Interesting Investigation: Can an SC-CNN-Based Circuit Behave Synchronously with the Original Chua's Circuit?

Since its introduction by Chua & Yang [25], cellular neural networks (CNNs) have attracted considerable interest, and many theoretical and experimental studies related to CNNs have been presented in the literature [4, 21, 41–42, 91, 96, 99, 113, 117, 123, 127, 133, 143]. One of these interesting studies has been reported by Arena *et al.* [7]. In their study, Arena *et al.* demonstrated that a suitable connection of three generalized CNN cells could generate the dynamics of Chua's circuit. This SC-CNN-based circuit, which consists of three generalized CNN cells, has been used as the chaos generator in CNN-based secure communications applications [16–17]. In the first part of this section, we perform the complete verification of the continuous synchronization phenomenon between two SC-CNN-based circuits using Pecora & Carroll's drive-response synchronization method [112]. After performing a complete verification of the continuous synchronization between two SC-CNN-based circuits depending on the driving variable, we discuss the synchronization between SC-CNN-based and Chua's circuit in the second part of this section.

6.3.1 *SC-CNN-based circuit*

Because it is the most preferable chaos generator, diverse realizations of Chua's circuit have been used in several applications. Arena *et al.* [7] have also derived a different version of Chua's circuit from a suitable connection of three simple generalized cellular neural network cells. Such a CNN has been called a state-controlled CNN by Arena *et al.* While Chua's circuit is governed by the following state equations in dimensionless form,

$$\dot{x} = \alpha[y - h(x)]$$
$$\dot{y} = x - y + z \qquad\qquad (6.15)$$
$$\dot{z} = -\beta y - \gamma z$$

where

$$h(x) = m_1 x + 0.5 \cdot (m_0 - m_1) \times (|x+1| - |x-1|) \qquad (6.16)$$

the SC-CNN-based circuit proposed by Arena *et al.* is defined by the following nonlinear state equations:

$$\dot{x}_1 = -x_1 + a_1 y_1 + s_{11} x_1 + s_{12} x_2$$
$$\dot{x}_2 = -x_2 + s_{21} x_1 + s_{23} x_3 \qquad\qquad (6.17)$$
$$\dot{x}_3 = -x_3 + s_{32} x_2 + s_{33} x_3$$

By determining "a" and "s" parameters in Eq. (6.17) as $a_1 = \alpha(m_1 - m_0)$; $s_{33} = 1-\gamma$; $s_{21} = s_{23} = 1$; $s_{11} = 1-\alpha \cdot m_1$; $s_{12} = \alpha$; and $s_{32} = -\beta$, Arena *et al.* showed that Chua's circuit defined by Eq. (6.15) can be obtained with x_1, x_2 and x_3, respectively, equal to x, y and z.

   The full circuit scheme of the SC-CNN-based circuit, which consists of three generalized CNN cells, and the cell connection scheme are shown in Fig. 6.10.

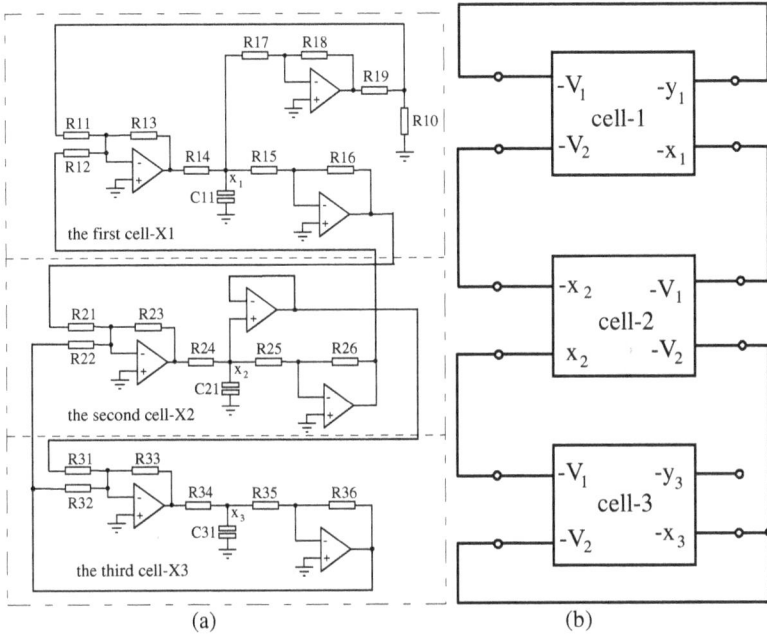

Fig. 6.10   (a) The full circuit scheme of the SC-CNN-based circuit, which consists of three generalized CNN cells, proposed by Arena *et al.* [7], (b) the cell connection scheme.

   In this circuit realization, circuit parameters were determined as $a_1 = 3.857$, $s_{11} = -1.5714$, $s_{32} = -14.286$, $s_{12} = 9$, $s_{21} = s_{23} = s_{33} = 1$ according to $a_1 = \alpha(m_1 - m_0)$; $s_{33} = 1 - \gamma$; $s_{21} = s_{23} = 1$; $s_{11} = 1 - \alpha \cdot m_1$; $s_{12} = \alpha$; and $s_{32} = -\beta$, such that the circuit structure exhibits well-known double-scroll chaotic attractor as in the original Chua's circuit with the parameter values $\beta = 14.286$, $\alpha = 9$, $\gamma = 0$, $m_0 = -1/7$, and $m_1 = 2/7$. From these design considerations defined by Arena *et al.* [7], we used the component values listed below for the SC-CNN–based circuit in Fig. 6.10(a): $R_{11} = 13.2$ kΩ, $R_{12} = 5.7$ kΩ, $R_{13} = 20$ kΩ, $R_{14} = 390$ Ω, $R_{15} = 100$ kΩ, $R_{16} = 100$ kΩ, $R_{17} = 74.8$ kΩ, $R_{18} = 970$ kΩ, $R_{19} = 27$ kΩ, $R_{10} = 2.22$ kΩ, $C_{11} = 51$ nF for the first cell, $R_{21} = R_{22} = R_{23} = R_{25} = R_{26} = 100$

kΩ, R_{24} = 1 kΩ, C_{21} = 51nF for the second cell, R_{32} = R_{33} = R_{35} = R_{36} = 100 kΩ, R_{34} = 1 kΩ, R_{31} = 7.8 kΩ, C_{31} = 51 nF for the third cell, and AD712 type voltage op amp with ±15 V supply voltages was used as an active element. Details of the design considerations of the realization of the SC-CNN-based circuit can be found in [7, 91].

6.3.2 *Continuous synchronization of SC-CNN-based circuits*

In this section, we present the results of the complete verification of chaos synchronization between two SC-CNN-based circuits [63], depending on the driving variable. To synchronize two SC-CNN-based circuits continuously, we used the Pecora-Carroll drive-response synchronization method, which is widely used for chaos synchronization. As is well known, in the drive-response method, Pecora-Carroll proposed constructing an identical copy of the response subsystem and driving it with the driving variables from the original system by dividing an n-dimensional system into two parts as the driving and response subsystems in an arbitrary way [112]. This drive-response method has been successfully applied to several systems, including Chua's circuit [23] and the Lorenz system [30]. In the literature, the drive-response approach has been applied to Chua's circuit in three configurations as x-drive, y-drive and z-drive configurations, and while it has been confirmed in x-drive and y-drive configurations, the subsystems have not synchronized in z-drive configurations [23].

As in Chua's circuit, we investigate whether chaos synchronization is achieved or not between two SC-CNN-based circuits according to x_1-drive, x_2-drive and x_3-drive configurations.

x_1-drive configuration: Fig. 6.11 shows the x_1-drive configuration between two identical SC-CNN-based circuits which exhibit well-known double-scroll attractor. The nonlinear state equations become for the drive system

$$\dot{x}_1 = -x_1 + a_1 y_1 + s_{11} x_1 + s_{12} x_2$$

$$\dot{x}_2 = -x_2 + s_{21} x_1 + s_{23} x_3 \qquad (6.18)$$

$$\dot{x}_3 = -x_3 + s_{32}x_2 + s_{33}x_3$$

and for the response system

$$\dot{x}_2' = -x_2' + s_{21}x_1 + s_{23}x_3'$$

$$\dot{x}_3' = -x_3' + s_{32}x_2' + s_{33}x_3' \qquad (6.19)$$

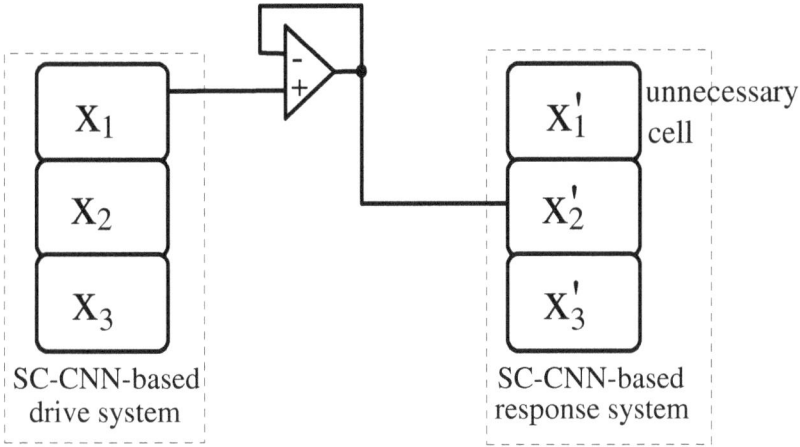

Fig. 6.11 x_1-drive configuration in SC-CNN-based system.

As shown in Fig. 6.11, x_1 chaotic dynamic of the first cell in the original system is chosen as the driving signal. Fig. 6.12 illustrates the synchronization graphs obtained between the drive and response systems. The double-scroll chaotic attractors exhibited in each system are shown in Fig. 6.13. These illustrations confirm that chaos synchronization is perfectly achieved between two SC-CNN-based circuits in x_1-drive configuration.

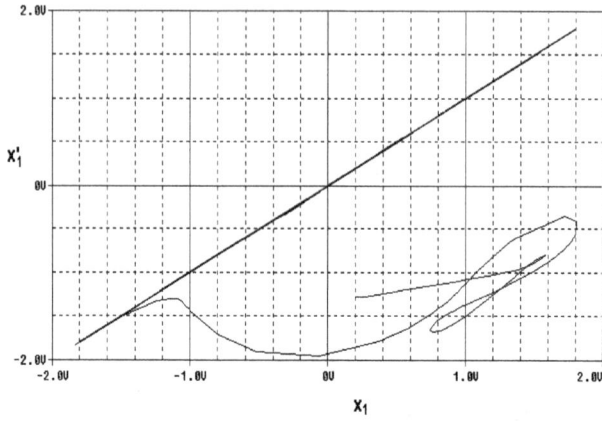

(a)

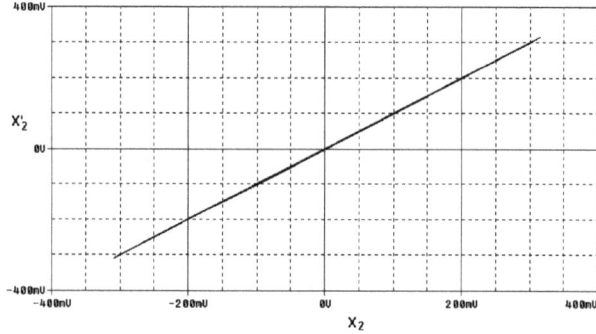

(b)

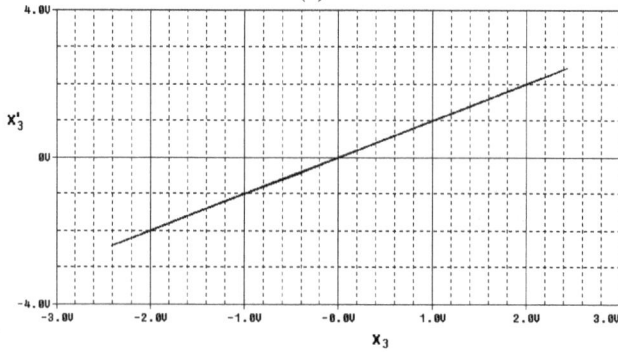

(c)

Fig. 6.12   The synchronization graphs between the cells of the drive and response systems in Fig. 6.11, (a) projection in x_1-x'_1 plane, (b) projection in x_2-x'_2 plane, (c) projection in x_3-x'_3 plane.

(a)

(b)

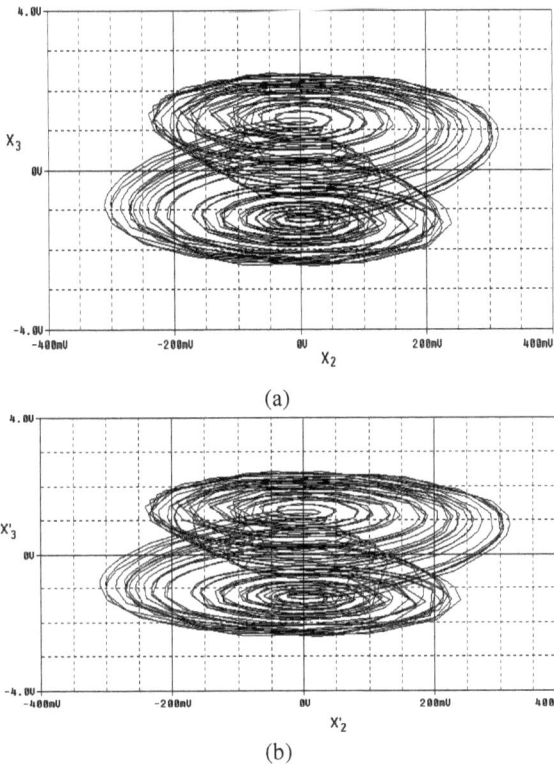

Fig. 6.13 (a) The double-scroll chaotic attractor exhibited in the drive system in Fig. 6.11, projection in x_2-x_3 plane, (b) the double-scroll chaotic attractor exhibited in the response system in Fig. 6.11, projection in x'_2-x'_3 plane.

x_2-drive configuration: Fig. 6.14 shows the x_2-drive configuration between two identical SC-CNN-based circuits. The nonlinear state equations for the drive system are

$$\dot{x}_1 = -x_1 + a_1 y_1 + s_{11} x_1 + s_{12} x_2$$

$$\dot{x}_2 = -x_2 + s_{21} x_1 + s_{23} x_3 \qquad (6.20)$$

$$\dot{x}_3 = -x_3 + s_{32} x_2 + s_{33} x_3$$

and for the response system

$$\dot{x}_1' = -x_1' + a_1 y_1' + s_{11} x_1' + s_{12} x_2$$

$$\dot{x}_3' = -x_3' + s_{32} x_2' + s_{33} x_3' \tag{6.21}$$

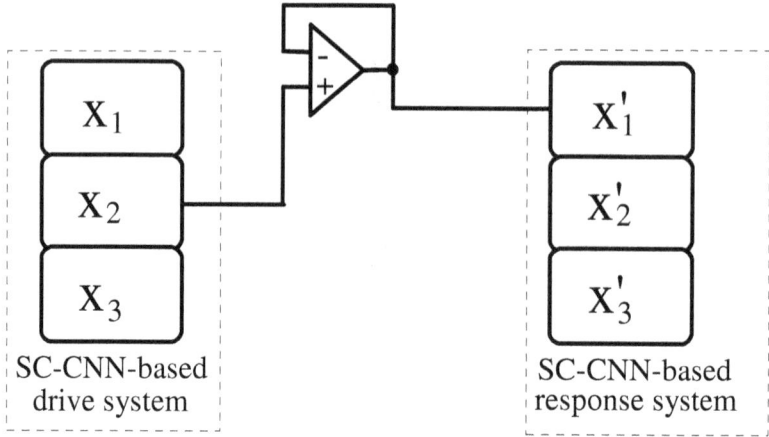

Fig. 6.14 x_2-drive configuration in SC-CNN-based system.

In this configuration, x_2 chaotic dynamic of the second cell in the original system is chosen as the driving signal. Fig. 6.15 illustrates the synchronization graphs obtained between the drive and response systems. The double-scroll chaotic attractors exhibited in each system are shown in Fig. 6.16. As shown in these figures, although chaos synchronization is also achieved in this x_2-drive configuration, the quality of the synchronization is not excellent as in the x_1-drive configuration.

It was confirmed in the literature that the subsystems of Chua's circuit have not been synchronized in the z-drive configuration. Similarly, the cells of the SC-CNN-based circuits don't synchronize in the x_3-drive configuration.

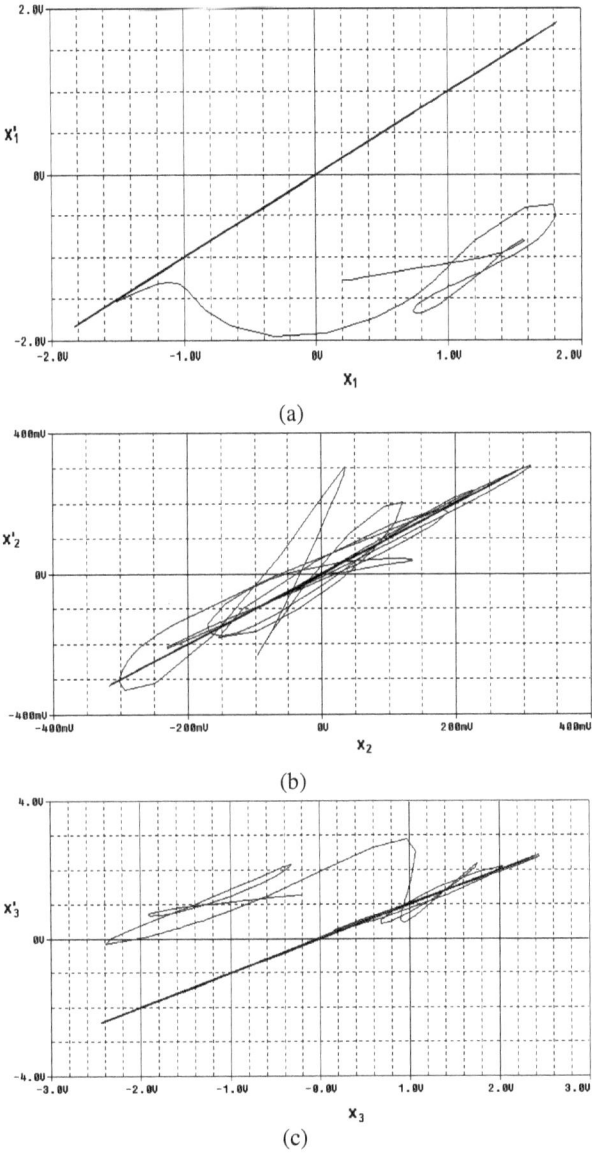

(a)

(b)

(c)

Fig. 6.15 The synchronization graphs between the cells of the drive and response systems in Fig. 6.14, (a) projection in x_1-x'_1 plane, (b) projection in x_2-x'_2 plane, (c) projection in x_3-x'_3 plane.

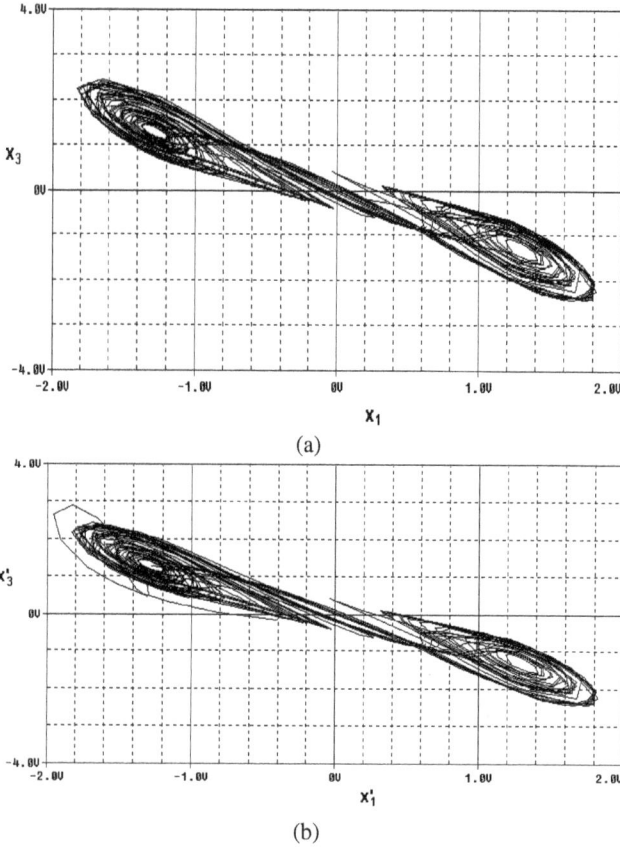

Fig. 6.16 (a) The double-scroll chaotic attractor exhibited in the drive system in Fig. 6.14, projection in x_1-x_3 plane, (b) the double-scroll chaotic attractor exhibited in the response system in Fig. 6.14, projection in x'_1-x'_3 plane.

6.3.3 *Can an SC-CNN-based circuit behave synchronously with the original Chua's circuit?*

For this investigation, we established a circuit scheme [63] in which a SC-CNN-based circuit is the drive system in one side and Chua's circuit is the response system in the other side. We determined the component values of Chua's circuit as $C_1 = 5.6$ nF, $C_2 = 50$ nF, $L = 7.14$ mH, and R = 1428 Ω yielding $\beta = 14.286$, $\alpha = 9$, $\gamma = 0$, and we used Kennedy's op-amp-based nonlinear resistor [57] in Chua's circuit yielding $m_0 = -1/7$

and $m_1 = 2/7$. We used the component values listed in Section 6.3.1 for the SC-CNN-based circuit as in Arena *et al.* [7], yielding $a_1 = 3.857$, $s_{11} = -1.5714$, $s_{32} = -14.286$, $s_{12} = 9$, $s_{21} = s_{23} = s_{33} = 1$ according to $a_1 = \alpha(m_1 - m_0)$; $s_{33} = 1 - \gamma$; $s_{21} = s_{23} = 1$; $s_{11} = 1 - \alpha \cdot m_1$; $s_{12} = \alpha$; and $s_{32} = -\beta$ such that the circuit structure exhibits the well-known double-scroll chaotic attractor as in the original Chua's circuit with the parameter values $\beta = 14.286$, $\alpha = 9$, $\gamma = 0$, $m_0 = -1/7$ and $m_1 = 2/7$.

It is worth noting that although the circuit implementations of the SC–CNN-based circuit and Chua's circuit are quite different, their parameter equalities are provided as much as possible. Now, we investigate this synchronization scheme using the Pecora-Carroll drive-response system.

x_1-**drive configuration:** Fig. 6.17 shows the x_1-drive configuration between the SC-CNN-based circuit and Chua's circuit.

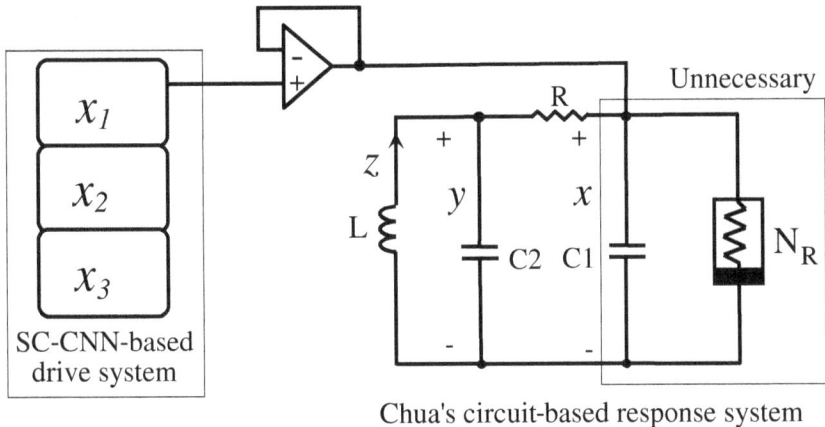

Fig. 6.17 The synchronization scheme formed in x_1-drive configuration between SC-CNN-based circuit and Chua's circuit.

The nonlinear state equations become for the drive system

$$\dot{x}_1 = -x_1 + a_1 y_1 + s_{11} x_1 + s_{12} x_2$$

$$\dot{x}_2 = -x_2 + s_{21} x_1 + s_{23} x_3 \qquad (6.22)$$

$$\dot{x}_3 = -x_3 + s_{32}x_2 + s_{33}x_3$$

and for the response system

$$\dot{y} = x_1 - y + z$$

$$\dot{z} = -\beta y - \gamma z \qquad (6.23)$$

As shown in Fig. 6.17, x_1 chaotic dynamic of the first cell of the SC-CNN-based circuit in the original system is chosen as the driving signal. And the subsystems of Chua's circuit are configured with respect to the Pecora-Carroll drive-response method. The chaotic dynamics of the SC-CNN-based drive system and Chua's circuit-based response system are shown in Fig. 6.18.

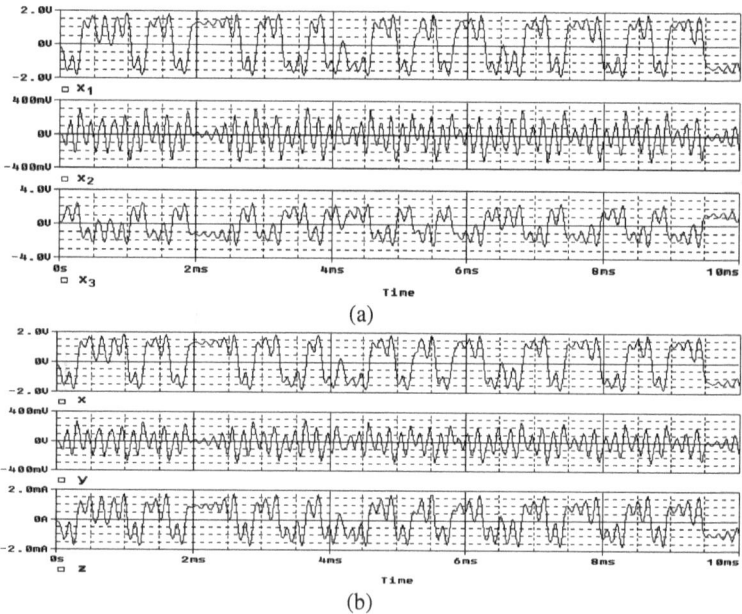

Fig. 6.18 For x_1-drive configuration in Fig. 6.17, (a) chaotic dynamics of the SC-CNN-based drive system, (b) chaotic dynamics of Chua's circuit-based response system.

Fig. 6.19 illustrates the synchronization graphs between the cell dynamics of the SC-CNN-based circuit and the subsystem dynamics of Chua's circuit. The double-scroll chaotic attractors exhibited in each system are shown in Fig. 6.20. These results show that chaos synchronization is achieved in sufficiently good quality with x_1-drive configuration and a SC-CNN-based circuit can behave synchronously with the original Chua's circuit in the case of very accurate parameter equalities.

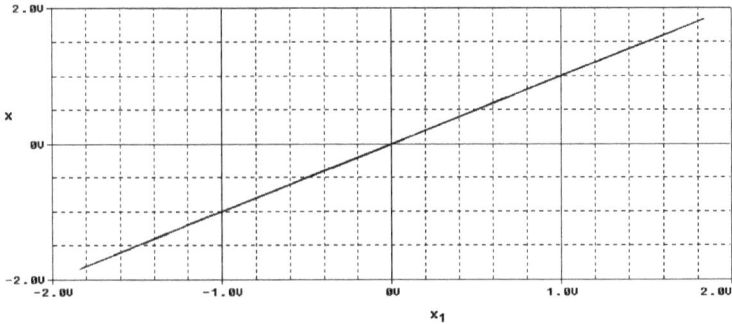

(a)

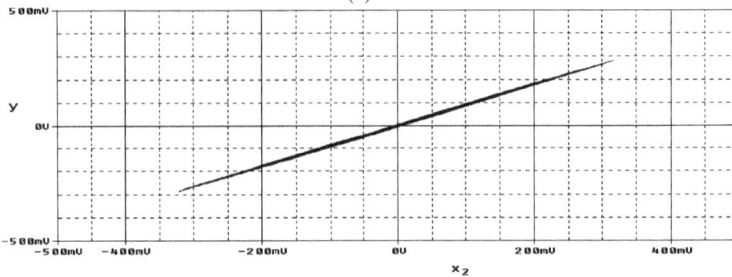

(b)

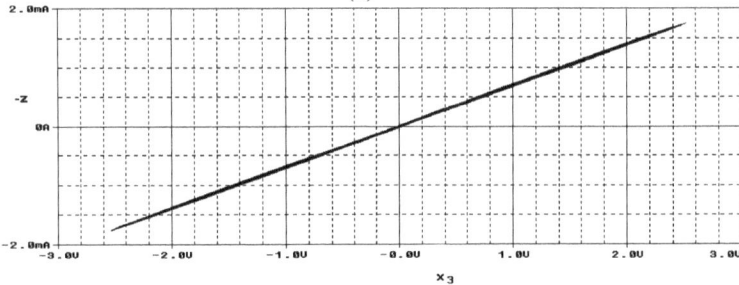

(c)

Fig. 6.19 The synchronization graphs between the drive and response systems in Fig. 6.17, (a) projection in x_1-x plane, (b) projection in x_2-y plane, (c) projection in x_3-(-z) plane.

(a)

(b)

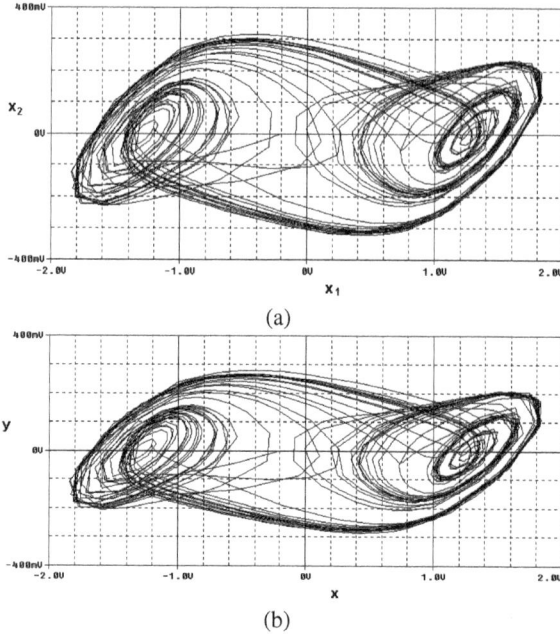

Fig. 6.20 (a) The double-scroll chaotic attractor exhibited in the drive system in Fig. 6.17, projection in x_1-x_2 plane, (b) the double-scroll chaotic attractor exhibited in the response system in Fig. 6.17, projection in x-y plane.

x_2-drive configuration: Fig. 6.21 shows the x_2-drive configuration between the SC-CNN-based circuit and Chua's circuit. The nonlinear state equations for the drive system are

$$\dot{x}_1 = -x_1 + a_1 y_1 + s_{11} x_1 + s_{12} x_2$$
$$\dot{x}_2 = -x_2 + s_{21} x_1 + s_{23} x_3 \qquad (6.24)$$
$$\dot{x}_3 = -x_3 + s_{32} x_2 + s_{33} x_3$$

and for the response system

$$\dot{x} = \alpha[x_2 - h(x)]$$
$$\dot{z} = -\beta y - \gamma z \qquad (6.25)$$

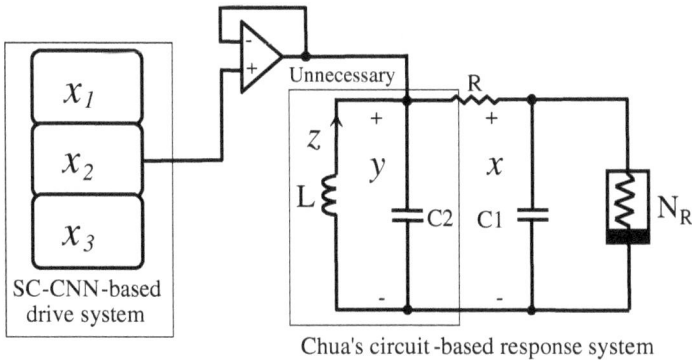

Fig. 6.21 The synchronization scheme between an SC-CNN-based circuit and Chua's circuit formed in x_2-drive configuration.

As shown in Fig. 6.21, x_2 chaotic dynamic of the second cell of the SC-CNN-based circuit in the original system is chosen as the driving signal. And the subsystems of Chua's circuit are configured with respect to the Pecora-Carroll drive-response method. Fig. 6.22 illustrates the synchronization graphs between the cell dynamics of the SC-CNN-based circuit and the subsystem dynamics of Chua's circuit. These results show that the quality of synchronization is poor in the proposed synchronization scheme formed by x_2-drive configuration.

6.4 Chaotic Masking System with Feedback Algorithm via SC-CNN-Based Circuit

In the literature, one of the most common chaotic communication schemes using chaos synchronization is the chaotic masking method. Chaos masking–based communication systems are generally implemented by using a Lorenz-based circuit [30] or Chua's circuit [85]. As shown in Fig. 6.23(a), in early proposed chaotic masking schemes, an output x(t) of a chaotic circuit such as Chua's circuit is added at the transmitter to an information signal of much less power s(t). The sum signal m(t) is transmitted to the receiver module in which an identical chaotic system tries to synchronize with x(t). After achieving

synchronization between the transmitter and receiver, the information
signal is retrieved at the receiver by subtraction of the chaotic signal x'(t)
from the sum signal m(t).

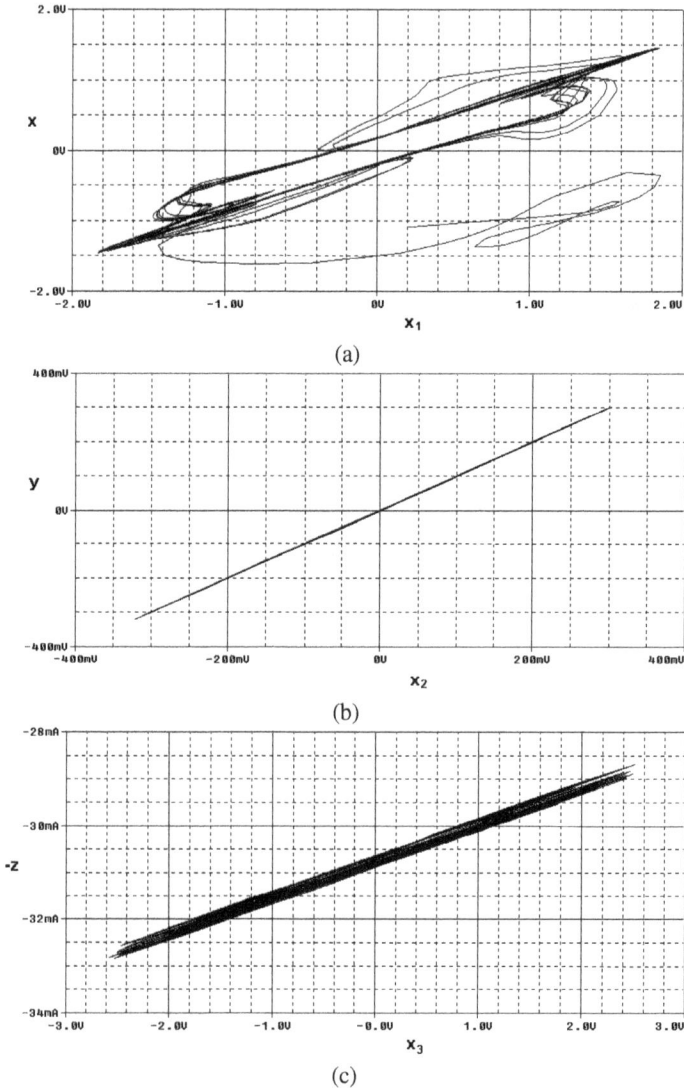

(a)

(b)

(c)

Fig. 6.22 The synchronization graphs between the drive and response systems in Fig.
6.21, (a) projection in x_1-x plane, (b) projection in x_2-y plane, (c) projection in x_3-(-z)
plane.

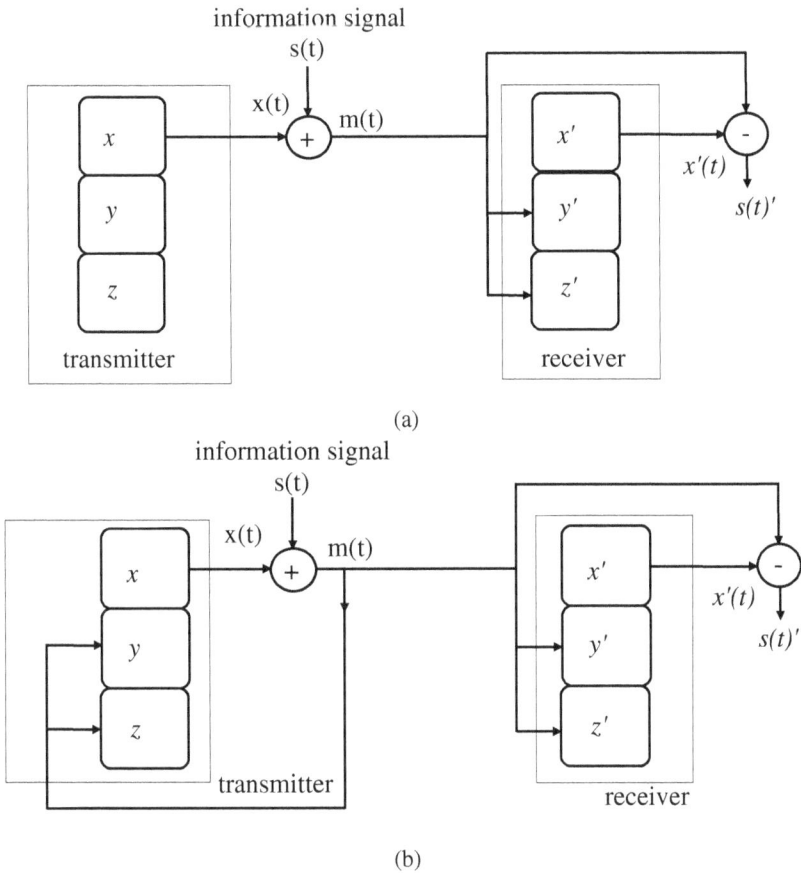

Fig. 6.23 (a) Chaotic masking system without feedback, (b) chaotic masking system with feedback algorithm.

Although the chaotic masking method is very simple, it has the following disadvantages:

a.   In order to keep chaos synchronization in a chaotic masking system, the level of the information signal must be significantly lower than that of the chaotic signal.

b.   If Chua's circuit is used as the chaos generator in a chaotic masking system, the demodulation is disrupted when the information signal approaches the natural frequency of the LC tank in Chua's circuit.

In order to avoid these problems and to improve the chaotic masking system, some algorithms have been proposed in the literature. One of these algorithms is the feedback algorithm [95]. In this algorithm, the transmitter is modified such that the combined signal m(t) is fed back into the system in addition to transmission as shown in Fig. 6.23(b). Since the systems are kept equivalent after the addition, the information signal s(t) then no longer affects the synchronization. This chaotic masking method with feedback was verified in Lorenz-based and Chua's circuit-based communication systems.

In this section, we show that a chaotic masking system with feedback can be constructed by the generalized SC-CNN cells [71]. For this purpose, we constructed a secure communication system based on chaotic masking with feedback method utilizing the generalized CNN cell model of Chua's circuit. The block diagram of the proposed SC-CNN-based chaotic masking system with feedback is shown in Fig. 6.24.

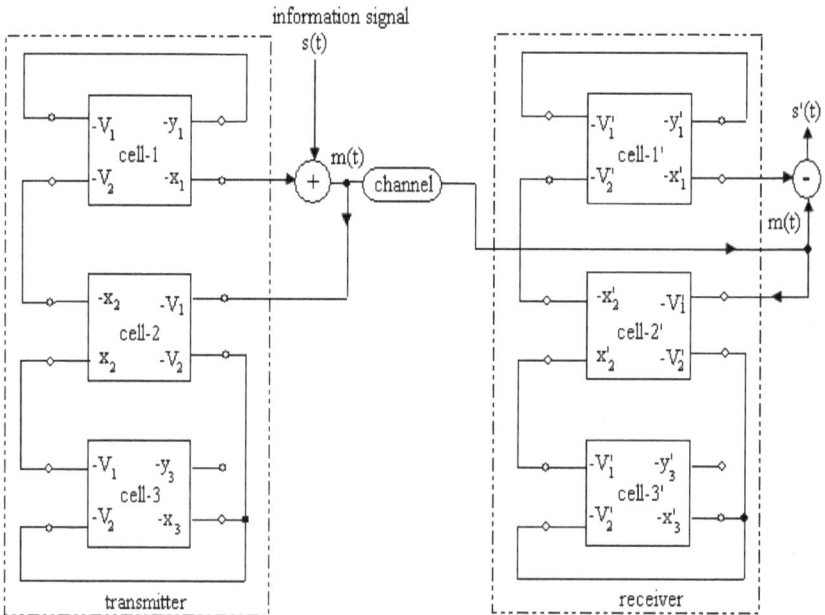

Fig. 6.24 The block diagram of the SC-CNN-based chaotic masking system with feedback algorithm.

As shown in Fig. 6.24, the first cell of the transmitter is the driver system, and the other two cells are the response subsystems.

Transmitter module without feedback is defined by the following state equations:

$$\dot{x}_1 = -x_1 + a_1 y_1 + s_{11} x_1 + s_{12} x_2$$

$$\dot{x}_2 = -x_2 + s_{21} x_1 + s_{23} x_3 \tag{6.26}$$

$$\dot{x}_3 = -x_3 + s_{32} x_2 + s_{33} x_3$$

In order to modify the transmitter, the sum signal $m(t) = x_1(t) + s(t)$ is fed into $(-V_1)$ input of the second cell by breaking the connection between $(-x_1)$ state variable of the first cell and $(-V_1)$ input of the second cell. In this case, modified transmitter is defined by

$$\dot{x}_1 = -x_1 + a_1 y_1 + s_{11} x_1 + s_{12} x_2$$

$$\dot{x}_2 = -x_2 + s_{21} m(t) + s_{23} x_3 \tag{6.27}$$

$$\dot{x}_3 = -x_3 + s_{32} x_2 + s_{33} x_3$$

Similarly, the receiver is defined by

$$\dot{x}_1' = -x_1' + a_1 y_1' + s_{11} x_1' + s_{12} x_2'$$

$$\dot{x}_2' = -x_2' + s_{21} m(t) + s_{23} x_3' \tag{6.28}$$

$$\dot{x}_3' = -x_3' + s_{32} x_2' + s_{33} x_3'$$

In these equations, the cell parameters of the transmitter and receiver modules were determined as $a_1 = 3.857$, $s_{11} = -1.5714$, $s_{32} = -14.286$, $s_{12} = 9$, $s_{21} = s_{23} = s_{33} = 1$ according to $a_1 = \alpha(m_1 - m_0)$; $s_{33} = 1 - \gamma$; $s_{11} = 1 - \alpha \cdot m_1$; $s_{12} = \alpha$, and $s_{32} = -\beta$ such that the circuit structure exhibits double-scroll chaotic attractor as in the original Chua's circuit with the parameter values $\beta = 14.286$, $\alpha = 9$, $\gamma = 0$, $m_0 = -1/7$ and $m_1 = 2/7$.

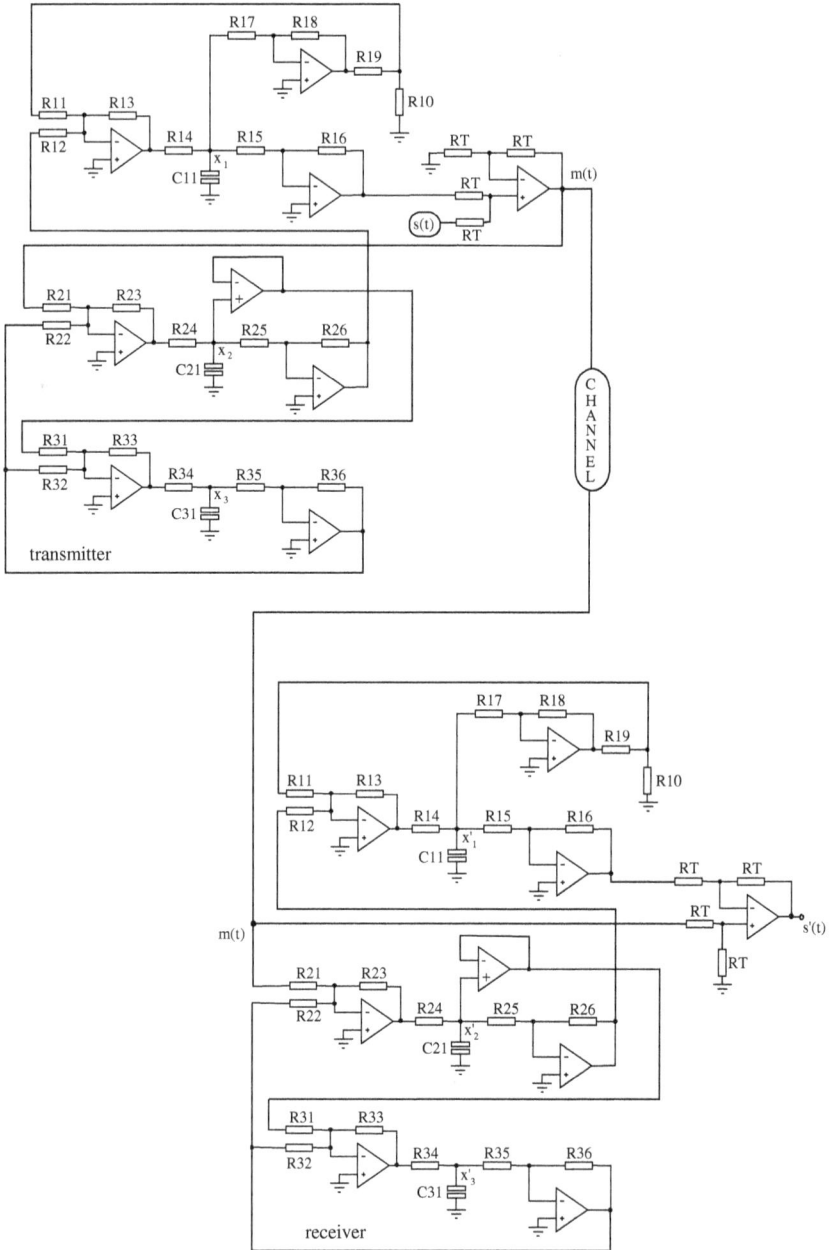

Fig. 6.25　The full circuit scheme of the SC-CNN-based chaotic masking system with feedback algorithm.

From these design considerations defined in Arena *et al.* [7], we determined the parameter values of the SC-CNN-based chaotic masking system with feedback, the full circuit scheme of which is shown in Fig. 6.25, as $R_{11} = 13.2$ kΩ, $R_{12} = 5.7$ kΩ, $R_{13} = 20$ kΩ, $R_{14} = 390$ Ω, $R_{15} = 100$ kΩ, $R_{16} = 100$ kΩ, $R_{17} = 74.8$ kΩ, $R_{18} = 970$ kΩ, $R_{19} = 27$ kΩ, $R_{10} = 2.22$ kΩ, $C_{11} = 51$ nF for the first cell; $R_{21} = R_{22} = R_{23} = R_{25} = R_{26} = 100$ kΩ, $R_{24} = 1$ kΩ, $C_{21} = 51$nF for the second cell; $R_{32} = R_{33} = R_{35} = R_{36} = 100$ kΩ, $R_{34} = 1$ kΩ, $R_{31} = 7.8$ kΩ, $C_{31} = 51$ nF for the third cell; and $R_T = 100$ kΩ for summing and subtracting circuits. In this realization, AD712 type voltage op amp with ± 15 V supply voltages was used as an active element.

6.4.1 *Simulation results*

To illustrate the synchronization performance of the chaotic masking system with feedback in Fig. 6.24, we used different analog signals as the information signal for different amplitude and frequency values. We demonstrate that this SC-CNN-based chaotic masking system with feedback exhibits satisfactory performance for various levels of the information signal by comparing this system with a SC-CNN-based chaotic masking system without feedback in the same transmission conditions.

In our first simulation experiments, we used sine and triangle waves as information signals with f = 1 kHz and A = 1–2 V amplitude range. We specifically chose a high-level information signal to illustrate that chaos synchronization is not disrupted in transmission of high-amplitude information signals. Fig. 6.26 illustrates the synchronization graphs between the transmitter and receiver, and chaotic attractors exhibited in two systems. In these graphs, the 45^0 lines and chaotic attractors indicate that in both transmitter and receiver modules nearly the same chaotic behavior is generated, and nearly perfect synchronization is achieved and maintained. The chaotic dynamic of the SC-CNN-based circuit, transmitted signal with information signal, input and output signals are shown in Fig. 6.27 and Fig. 6.28 for sine wave and triangle wave inputs, respectively.

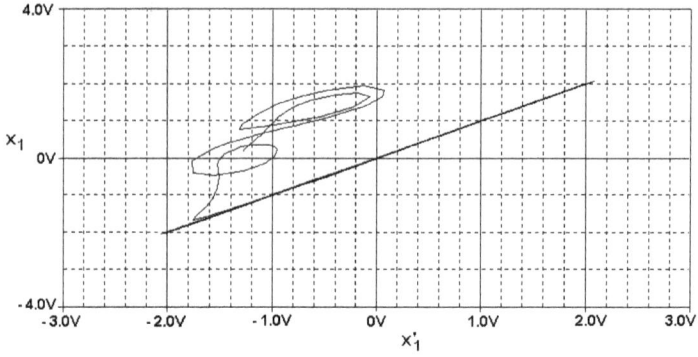

(a)

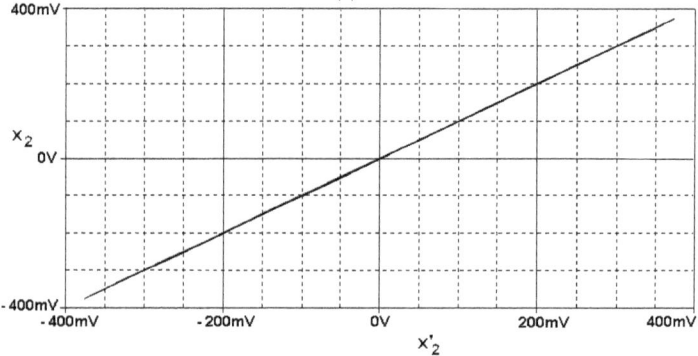

(b)

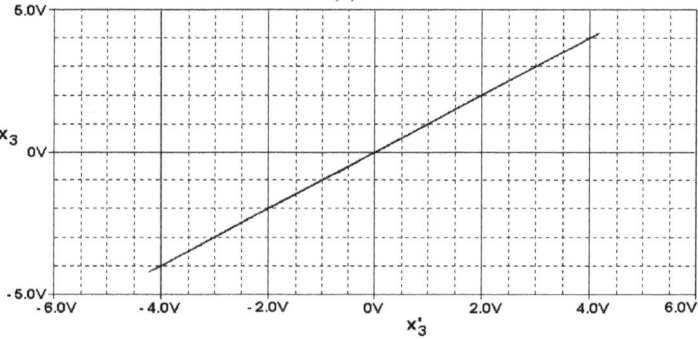

(c)

Fig. 6.26 (a) The synchronization graph between the first cells of the transmitter and receiver, projection in the x_1-x'_1 plane, (b) the synchronization graph between the second cells of the transmitter and receiver, projection in the x_2-x'_2 plane, (c) the synchronization graph between the third cells of the transmitter and receiver, projection in the x_3-x'_3 plane.

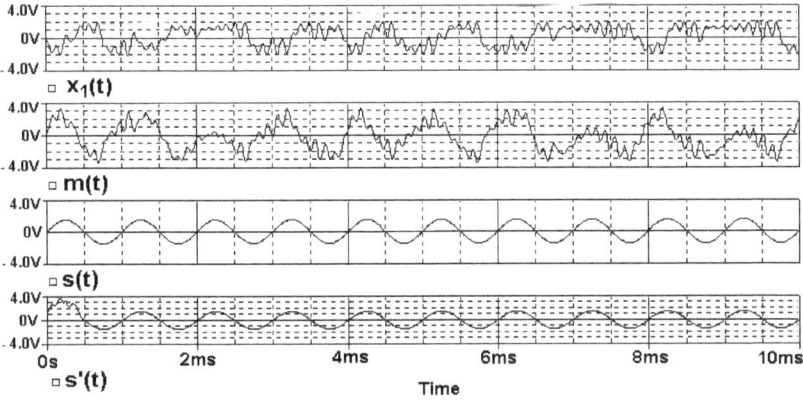

Fig. 6.27   (a) The chaotic signal $x_1(t)$ generated in the transmitter circuit, (b) the transmitted signal $m(t) = x_1(t) + s(t)$, (c) input signal s(t) of frequency 1 KHz in sinusoidal waveform, (d) recovered signal $s'(t)$ in the receiver.

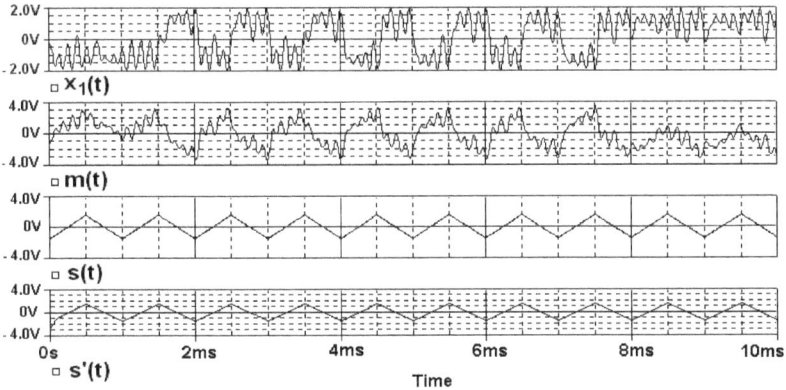

Fig. 6.28   (a) The chaotic signal $x_1(t)$ generated in the transmitter circuit, (b) the transmitted signal $m(t) = x_1(t) + s(t)$, (c) input signal s(t) of frequency 1 kHz in triangle waveform, (d) recovered signal $s'(t)$ in the receiver.

In order to demonstrate the effectiveness of this SC-CNN-based chaotic masking system using feedback, the same simulation experiments were performed by using the same transmission conditions in the chaotic masking system without feedback. For this purpose, we disconnected the feedback path in the chaotic masking system in Fig. 6.24. In this case, such a high level of information signal strongly affects the synchronization between the transmitter and receiver. For this operation mode, without feedback, while the synchronization graph is shown in Fig. 6.29, the chaotic dynamic of the transmitter, transmitted signal with information signal, input and output signals are shown in Fig. 6.30, respectively.

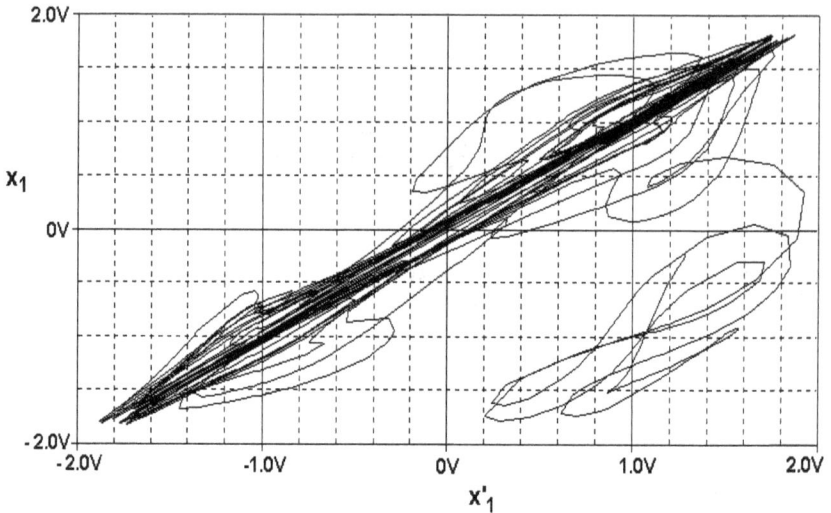

Fig. 6.29   The synchronization graph between the transmitter and receiver in the case when the feedback connection in the SC-CNN-based chaotic masking system in Fig. 6.25 is broken.

In the proposed chaotic masking system with feedback in Fig. 6.25, high-frequency chaotic oscillations can be obtained by adjusting the circuit parameters, and in this mode it is possible to transmit high-frequency information signals. By determining capacitors in the cells of the transmitter and receiver as $C_1 = C_2 = C_3 = 22$ nF, we reconfigured the communication system in Fig. 6.25, and we tried to transmit 100 kHz–1 MHz information signals. In this case, the chaotic attractor, synchronization graph, input and output signals are shown in Fig. 6.31, respectively.

As shown in the figure, after a transition period at the beginning, the synchronization is achieved, and high-frequency information signals can also be transmitted successfully in this SC-CNN-based communication system.

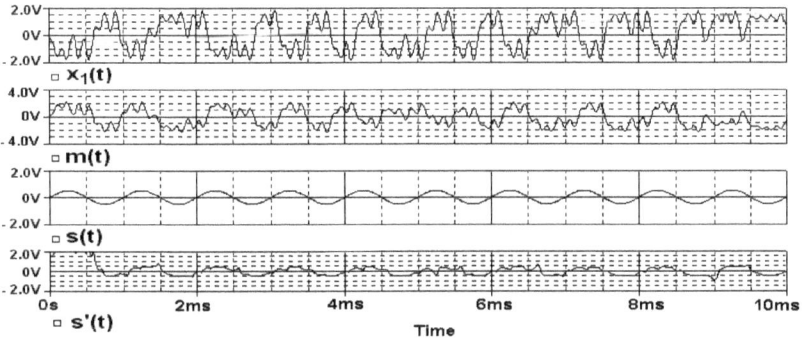

Fig. 6.30   The case where the feedback connection in SC-CNN-based chaotic masking system in Fig. 6.25 is broken, (a) the chaotic signal $x_1(t)$ generated in the transmitter circuit, (b) the transmitted signal m(t) = $x_1(t)$ + s(t), (c) input signal s(t) of frequency 1 kHz in sinusoidal waveform, (d) recovered signal s'(t) in the receiver.

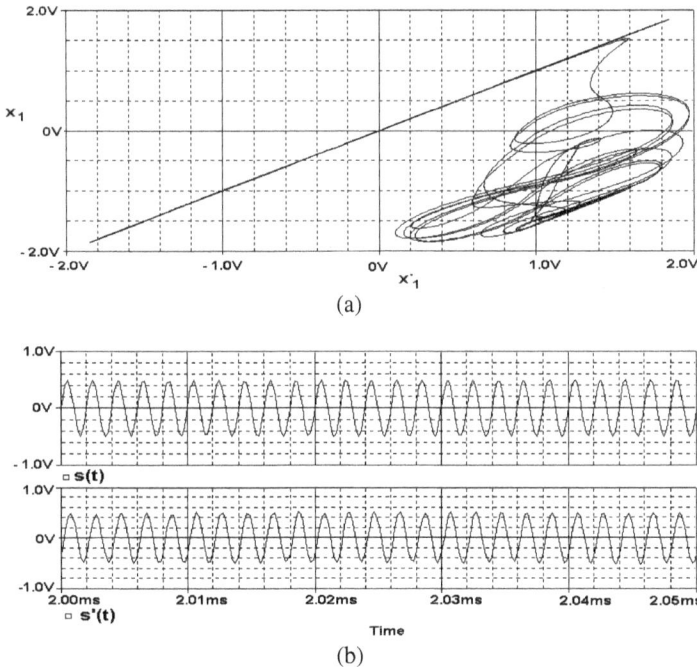

(a)

(b)

Fig. 6.31   For transmission of 500 kHz information signal in chaotic communication system in Fig. 6.25, (a) the synchronization graph, (b) input signal in sinusoidal waveform and recovered signal.

6.5 Impulsive Synchronization Studies Using SC-CNN-Based Circuit and Chua's Circuit

So far, the impulsive synchronization method has been applied to several well-known chaotic circuits and systems such as Chua's circuit [55, 111], Lorenz system [145] and a hyperchaotic circuit [55]. Panas *et al.* [111] presented the first experimental results on the impulsive synchronization by using two chaotic Chua's circuits. In their study, it has been shown that two Chua's circuits are synchronized impulsively by using narrow impulses. The more detailed studies of impulsive synchronization and its performance analysis have been reported in Itoh *et al.* [55] by using Chua's circuit and a hyperchaotic circuit, and its performance was verified by using some spread spectrum communication systems which use Chua's circuit and a hyperchaotic circuit as the chaos generator.

In this section, we considered applying the impulsive synchronization method to two separate synchronization schemes [72]. For this purpose, in the first part of this section, we chose an SC-CNN-based circuit as the chaotic circuit and investigated whether two SC-CNN-based circuits are synchronized impulsively or not by evaluating the minimum length of impulse width (Q) and the ratio of impulse width to impulse period (Q/T). In the second part of this section, we investigate the impulsive synchronization between an SC-CNN-based circuit and Chua's circuit. For this purpose, by choosing an SC-CNN-based circuit and Chua's circuit as driving and driven chaotic circuits, we investigated whether these circuits are synchronized impulsively or not by evaluating the minimum length of impulse width (Q) and the ratio of impulse width to impulse period (Q/T).

6.5.1 *Impulsive synchronization of chaotic circuits*

The studies related to the impulsive synchronization of chaotic circuits were generally realized by using Chua's circuit. In addition to Chua's circuit, the Lorenz system and a hyperchaotic circuit were also used in the impulsive synchronization schemes. The first experimental study on the impulsive synchronization was also realized by using Chua's circuit.

(a)

(b)

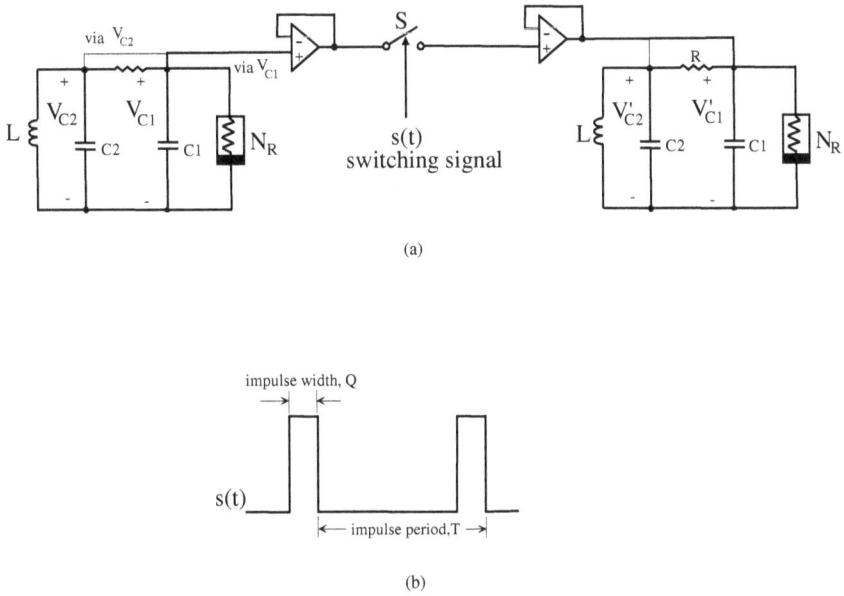

Fig. 6.32   (a) The block diagram of impulsive synchronization between two Chua's circuits, (b) the switching signal s(t) used to turn on/off a switching device (S) in Fig. 6.32(a).

The block diagram of the Chua's-circuit-based impulsive synchronization scheme is shown in Fig. 6.32(a). As shown in Fig. 6.32(a), this impulsive synchronization scheme uses a single synchronizing impulse sequence, and impulsive synchronization can be realized via V_{C1} or V_{C2} between two Chua's circuits. To synchronize two Chua's circuits impulsively, impulses sampled from one state variable (V_{C1} or V_{C2}) of the driving circuit to the driven circuit are transmitted through one communication channel. The switching signal s(t), shown in Fig. 6.32(b), which is used to turn on/off a switching device, consists of the impulse width Q and the impulse period T. While Q represents the width of the high-level peaks, T represents the time interval between two consecutive impulses.

In the first experimental study on impulsive synchronization [111], it was observed that two Chua's circuits are synchronized by using narrow impulses (Q = 8 µs, T = 50 µs, Q/T = 16%).

After this observation, by configuring the experimental set up of impulsive synchronization by using two Chua's circuits, Itoh *et al.* [55] have evaluated the minimum length of the interval Q which gives the perfect synchronization, and the ratio of Q to T. They have reported the following results from their experimental study:

I. For T ≤ 9 μs
In this case, only 8% ratio of Q/T is required for perfect synchronization.

II. For 9 μs ≤ T ≤ 50 μs
In Itoh *et al.* [55], for this case it is stated that the ratio Q/T increases from 8% to 50% as T increases.

III. For 50 μs ≤ T ≤ 5 ms
In this case, the ratio Q/T must be at least 50% to achieve an almost-identical synchronization.

Before beginning our study related on the impulsive synchronization of SC-CNN-based circuits, we referenced the above results, and by evaluating the minimum length of the interval Q and the ratio of Q to T, we also tried to constitute a similar synchronization criterion for impulsive synchronization of SC-CNN-based circuits. These results obtained from our simulation experiments are given in the following sections.

6.5.2 *Impulsive synchronization of SC-CNN-based circuits*

In this section, we present our simulation results on the impulsive synchronization between two SC-CNN-based circuits, and we report the minimum impulse width Q and the ratio Q/T for achieving impulsive synchronization under different impulse frequencies. We will also compare these simulation results of the impulsive synchronization between two SC-CNN-based circuits with the results of the impulsive synchronization between two Chua's circuits reported in Itoh *et al.* [55].

The impulsive synchronization between two SC-CNN-based circuits can be realized via two cell dynamics x_1, and x_2. So, we configured two impulsive synchronization schemes between two SC-CNN-based circuits [72]. The block diagrams of impulsive synchronization via two cell dynamics x_1, and x_2 between two SC-CNN-based circuits are shown in Fig. 6.33(a) and 6.33(b), respectively.

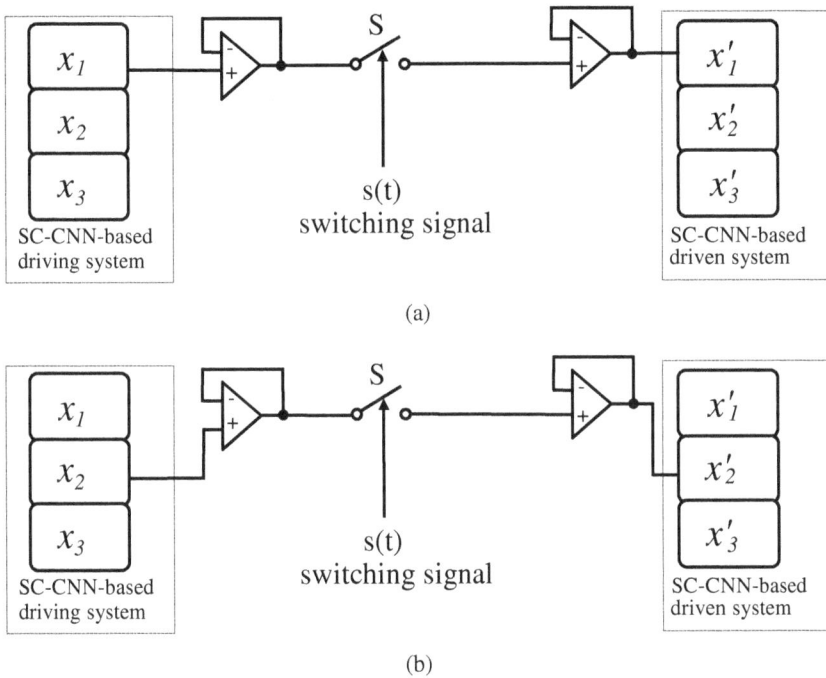

(a)

(b)

Fig. 6.33 The block diagrams of impulsive synchronization between two SC-CNN-based circuits, (a) impulsive synchronization via x_1, (b) impulsive synchronization via x_2.

6.5.2.1 *Impulsive synchronization via x_1 between two SC-CNN-based circuits*

The block diagram of impulsive synchronization via x_1 dynamic between two SC-CNN-based circuits is shown in Fig. 6.33(a). Here, x_1 dynamic of the driving circuit is chosen as the driving signal. The switch, S, is a voltage-controlled device.

When the switching signal $s(t)$ is at a high level, the switch S is "ON" and a synchronizing impulse is transmitted from the driving system to the driven system. When the switching signal s(t) is in the low level, the switch is turned off and two SC-CNN-based circuits are disconnected. As the minimum impulse width Q and the impulse period T can affect the performance of impulsive synchronization; in our study we investigated the system for different Q and T values. In this process, we referenced the results of impulsive synchronization between two Chua's circuits presented in Itoh *et al.* [55]. We observed that two SC-CNN-based circuits are synchronized impulsively for different Q and T values. As expected, if impulsive coupling is not applied, the two SC-CNN-based circuits are not synchronized. This desynchronization case is shown in Fig. 6.34, projection on the x_2-x'_2 plane. When impulsive coupling is applied, the two systems are synchronized impulsively. After investigating the impulsive synchronization via x_1 between two SC-CNN-based circuits for different Q and T values, we can draw the following conclusions from our simulation experiments similar to Itoh *et al.* [55]:

I. For T≤ 12 μs

In this case, two SC-CNN-based circuits can be synchronized impulsively with 5–10% Q/T ratio, and Q increases in proportion to the length of T. In Fig. 6.35(a)–(c), the synchronization states on the x_1-x'_1 plane, x_2-x'_2 plane and x_3-x'_3 plane are shown for Q = 0.8 μs and T = 10 μs, yielding ratio Q/T = 8%.

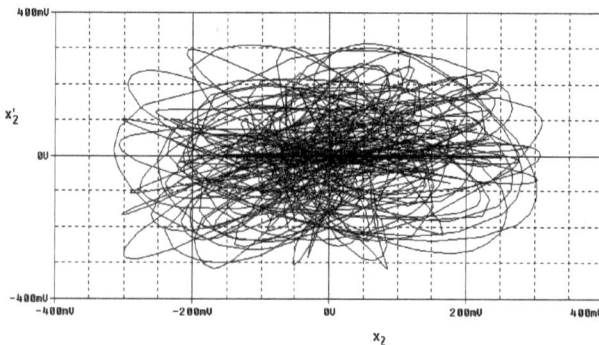

Fig. 6.34 The desynchronization graph between the driving and driven circuits in Fig. 6.33(a).

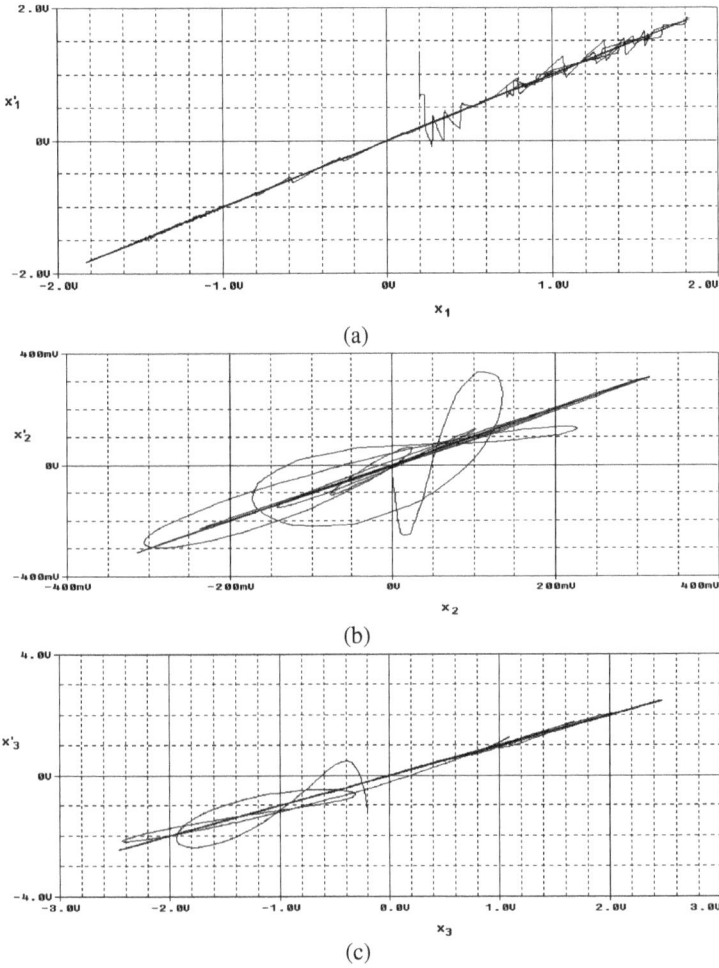

(a)

(b)

(c)

Fig. 6.35 For Q = 0.8 μs, T = 10 μs (Q/T = 8%), the synchronization graphs of the impulsive synchronization scheme via x_1, (a) projection in x_1-x'_1 plane, (b) projection in x_2-x'_2 plane, (c) projection in x_3-x'_3 plane.

II. For 12 μs ≤ T ≤ 45 μs

In this case, two SC-CNN-based circuits can be synchronized when the ratio Q/T increases from 5–10% to 50% as T increases. In Fig. 6.36(a)–(c), the synchronization states on the x_1-x'_1 plane, x_2-x'_2 plane and x_3-x'_3 plane are shown for Q = 12 μs and T = 30 μs, yielding ratio Q/T = 40%.

(a)

(b)

(c)

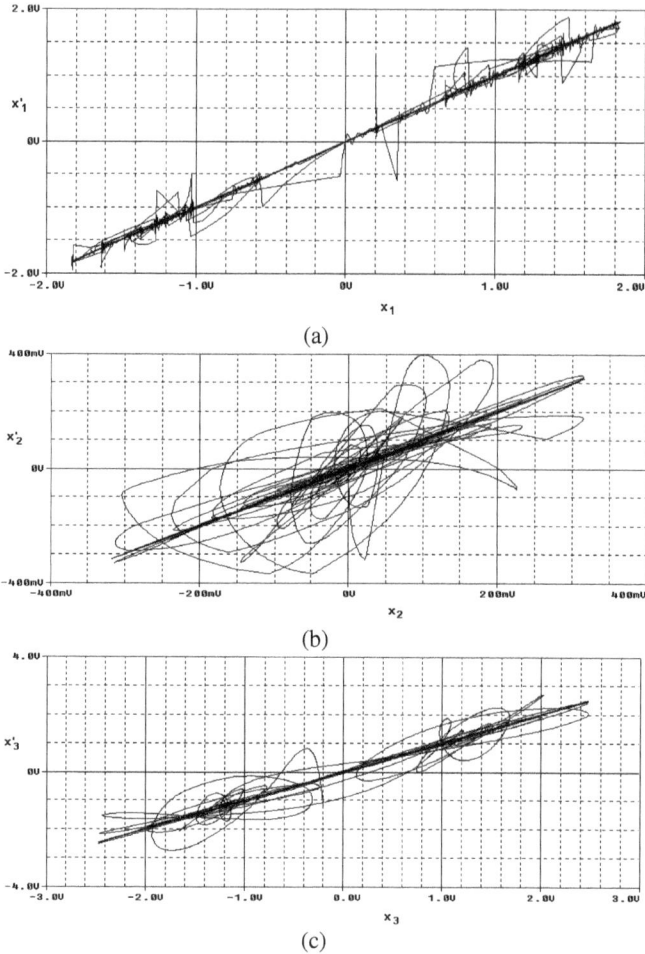

Fig. 6.36   For Q = 12 μs, T = 30 μs (Q/T = 40%), the synchronization graphs of the impulsive synchronization scheme via x_1, (a) projection in x_1-x'_1 plane, (b) projection in x_2-x'_2 plane, (c) projection in x_3-x'_3 plane.

III. For T ≥ 45 μs

In this case, to achieve synchronization impulsively, the impulse width occupies at least 50% of the impulse period. In Fig. 6.37(a)–(c), the synchronization states on the x_1-x'_1 plane, x_2-x'_2 plane and x_3-x'_3 plane are shown for Q = 36 μs and T = 60 μs, yielding ratio Q/T = 60%.

(a)

(b)

(c)

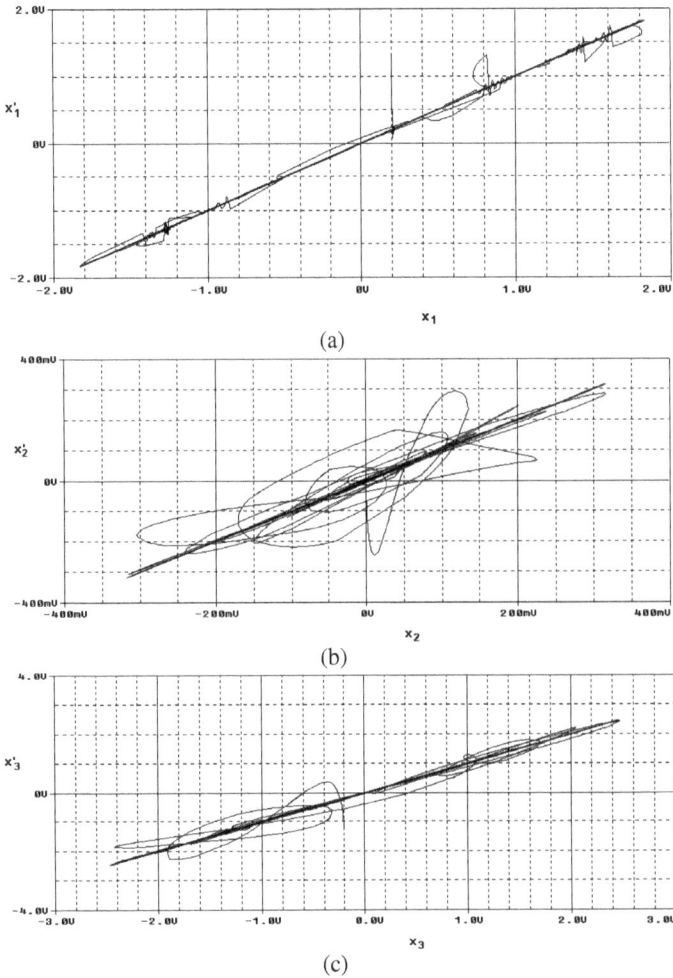

Fig. 6.37   For $Q = 36$ μs, $T = 60$ μs ($Q/T = 60\%$), the synchronization graphs of the impulsive synchronization scheme via x_1, (a) projection in x_1-x'_1 plane, (b) projection in x_2-x'_2 plane, (c) projection in x_3-x'_3 plane.

6.5.2.2   *Impulsive synchronization via x_2 between two SC-CNN-based circuits*

Figure 6.33(b) shows the block diagram of the impulse synchronization scheme via x_2 between two SC-CNN-based circuits. In this scheme, the second cell dynamics of both SC-CNN-based circuits are connected impulsively by a voltage buffer and a voltage-controlled switch. If impulsive coupling is not applied, the two SC-CNN-based circuits are not synchronized. This desynchronization case is shown in the Fig. 6.38 projection on the x_1-x'_1 plane.

When impulsive coupling is applied, two systems are synchronized impulsively for different Q and T values. After investigating the impulsive synchronization via x_2 between two SC-CNN-based circuits for different Q and T values, we can draw the following conclusions from our simulation experiments:

I. For T≤ 55 μs

In this case, two SC-CNN-based circuits can be synchronized impulsively with 5–10% Q/T ratio, and Q increases in proportion to the length of T. In Fig. 6.39 (a)–(c), the synchronization states on the x_1-x'_1 plane, x_2-x'_2 plane and x_3-x'_3 plane are shown for Q = 2 μs and T = 20 μs, yielding Q/T = 10%.

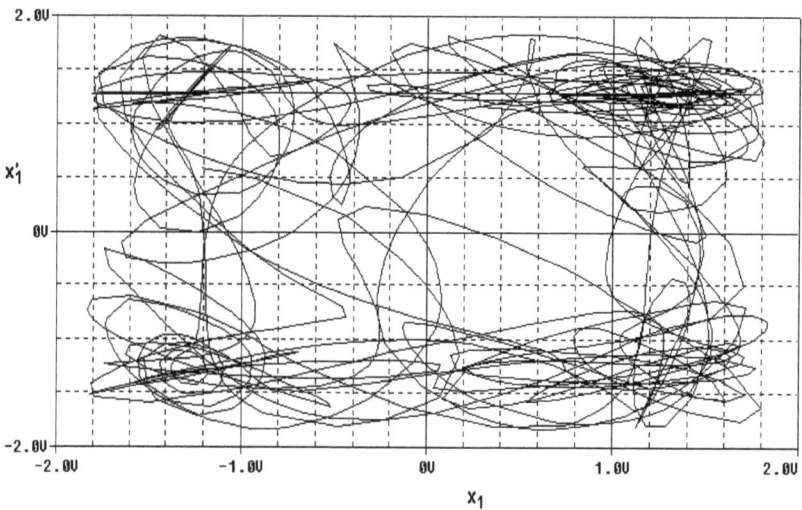

Fig. 6.38 Desynchronization graph between driving and driven circuits in Fig. 6.33(b).

(a)

(b)

(c)

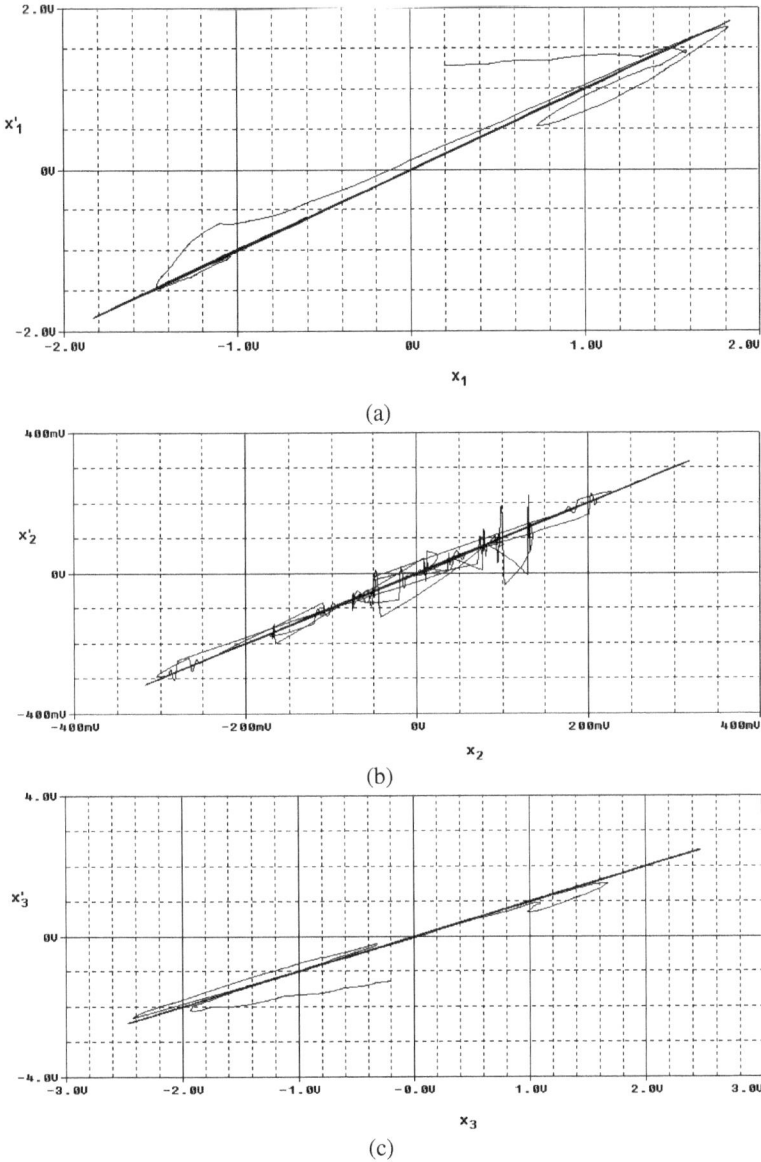

Fig. 6.39   For Q = 2 μs, T = 20 μs (Q/T = 10%), the synchronization graphs of the impulsive synchronization scheme via x_2, (a) projection in x_1-x'_1 plane, (b) projection in x_2-x'_2 plane, (c) projection in x_3-x'_3 plane.

II. For 55 µs ≤ T ≤70 µs

In this case, two SC-CNN-based circuits can be synchronized when the ratio Q/T increases from 30% to 50% as T increases. In Fig. 6.40(a)–(c), the synchronization states on the x_1-x'_1 plane, x_2-x'_2 plane and x_3-x'_3 plane are shown for Q = 18 µs and T = 60 µs, yielding Q/T = 30%.

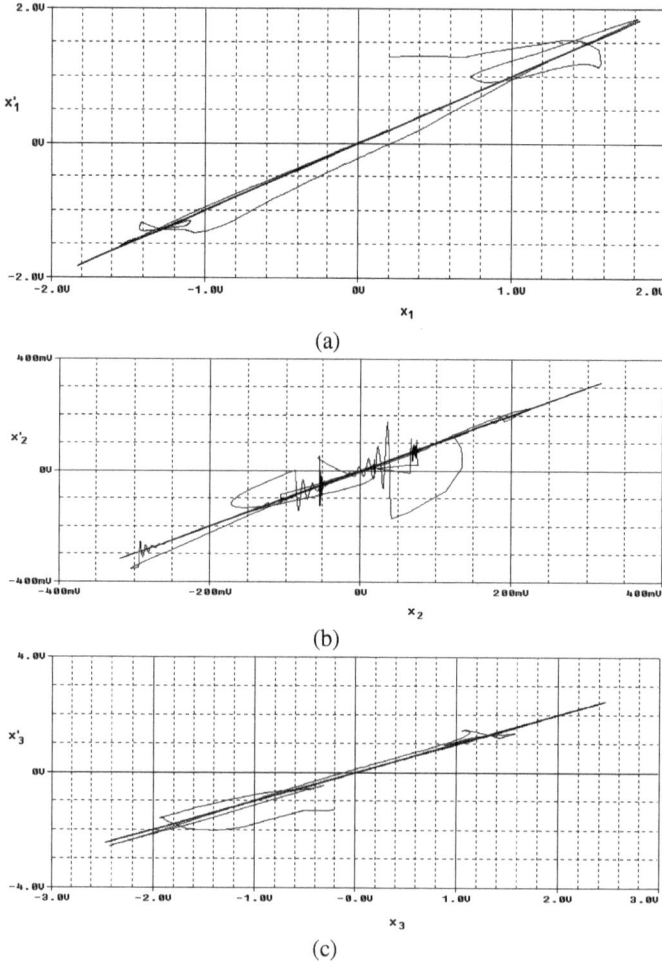

(a)

(b)

(c)

Fig. 6.40 For Q = 18 µs, T = 60 µs (Q/T = 30%), the synchronization graphs of the impulsive synchronization scheme via x_2, (a) projection in x_1-x'_1 plane, (b) projection in x_2-x'_2 plane, (c) projection in x_3-x'_3 plane.

III. For T ≥ 70 μs

In this case, to achieve synchronization impulsively, the impulse width occupies at least 50% of the impulse period. In Fig. 6.41(a)–(c), the synchronization states on the x_1-x'_1 plane, x_2-x'_2 plane and x_3-x'_3 plane are shown for Q = 48 μs and T = 80 μs, yielding Q/T = 60%.

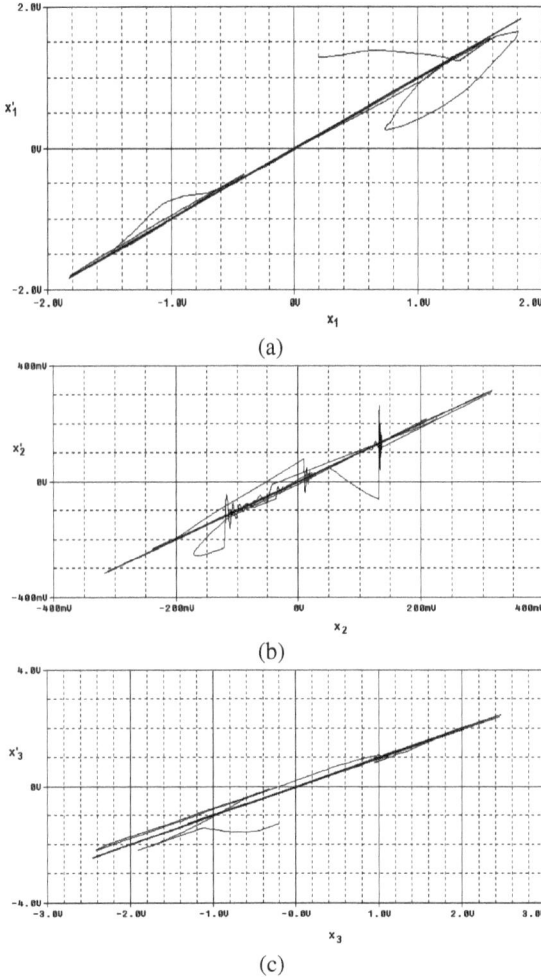

(a)

(b)

(c)

Fig. 6.41   For Q = 48 μs, T = 80 μs (Q/T = 60%), the synchronization graphs of the impulsive synchronization scheme via x_2, (a) projection in x_1-x'_1 plane, (b) projection in x_2-x'_2 plane, (c) projection in x_3-x'_3 plane.

We also investigated the impulsive synchronization via x_3 between two SC-CNN-based circuits. It was confirmed in the literature that the cells of the SC-CNN-based circuits don't synchronize in the x_3-drive configuration in continuous synchronization. Similar to the case of continuous synchronization, we observed that two SC-CNN-based circuits in the x_3-drive configuration don't synchronize impulsively.

6.5.3 *Impulsive synchronization between SC-CNN-based circuit and Chua's circuit*

A block diagram of impulsive synchronization between an SC-CNN-based circuit and Chua's circuit [72] is given in Fig. 6.42. As shown in the figure, an SC-CNN-based circuit is chosen as the driving circuit, and Chua's circuit is chosen as the driven circuit. This impulsive synchronization scheme uses a single synchronizing impulse sequence, and impulsive synchronization is realized via x_1-drive configuration. To synchronize the SC-CNN-based circuit and Chua's circuit impulsively, impulses sampled from cell dynamic (x_1) of the driving circuit to the driven circuit are transmitted through one communication channel.

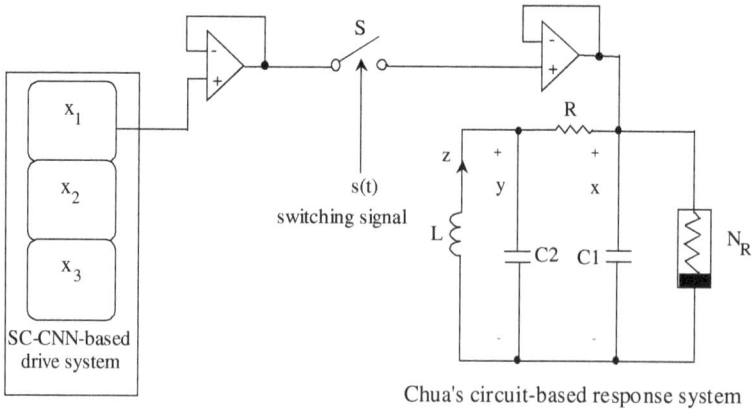

Fig. 6.42 The block diagram of impulsive synchronization between SC-CNN-based circuit and Chua's circuit via x_1.

When the switching signal s(t) is at a high level, the S-switch is "ON" and a synchronizing impulse is transmitted from the driving system to the driven system. When the switching signal s(t) is at a low level, the switch is turned off and the SC-CNN-based circuit and Chua's circuit are disconnected. As the minimum impulse width Q and the impulse period T can affect the performance of impulsive synchronization, in our study we investigated the system for different Q and T values. As expected, if impulsive coupling is not applied, the SC-CNN-based circuit and Chua's circuit are not synchronized. This desynchronization case is shown in Fig. 6.43, projection on the x_2-y plane. When impulsive coupling is applied, two systems are synchronized impulsively. After investigating the impulsive synchronization via x_1 between the SC-CNN-based circuit and Chua's circuit for different Q and T values, we can draw the following conclusions from our simulation experiments:

I. For T≤ 20 μs

In this case, an SC-CNN-based circuit and Chua's circuit can be synchronized impulsively with a 10% Q/T ratio, and Q increases in proportion to the length of T. The quality of the synchronization with 10% Q/T ratio is not sufficiently good. To improve the quality of synchronization in this case, the Q/T ratio should be increased to higher values.

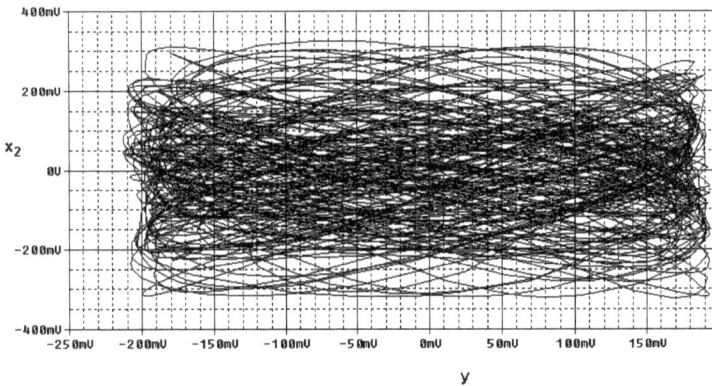

Fig. 6.43 Desynchronization graph between driving and driven circuits in Fig. 6.42.

In Fig. 6.44(a)–(b), the synchronization states on the x_1-x plane and x_2-y plane are shown for Q = 1.5 μs and T = 15 μs, yielding ratio Q/T = 10%.

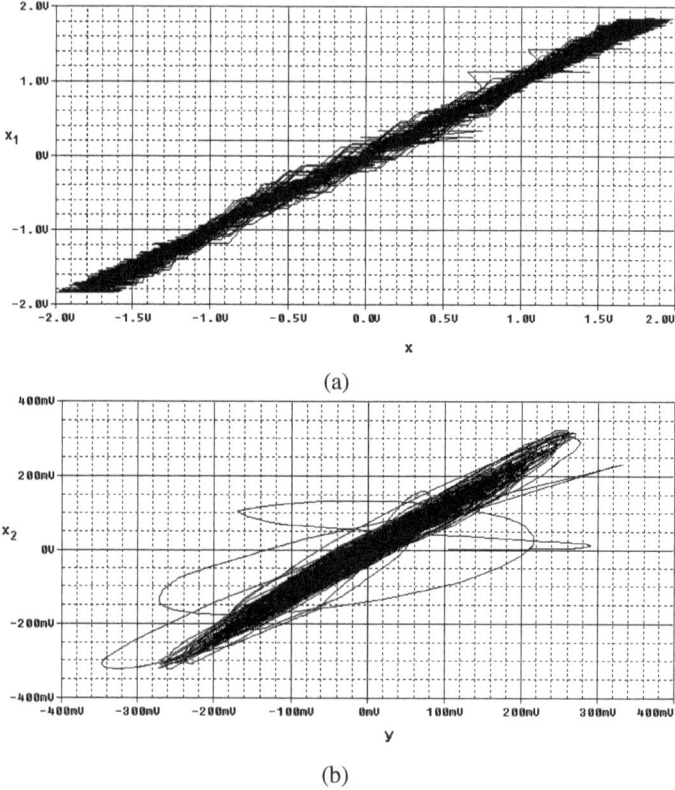

(a)

(b)

Fig. 6.44   For Q = 1.5 μs, T = 15 μs (Q/T = 10%), the synchronization graphs of the impulsive synchronization scheme in Fig. 6.42, (a) projection in x_1-x plane, (b) projection in x_2-y plane.

II. For 20 μs ≤ T ≤ 40 μs

In this case, an SC-CNN-based circuit and Chua's circuit can be synchronized when the ratio Q/T increases from 10% to 50% as T increases. Q/T ratios near 50% provide better-quality synchronization cases. In Fig. 6.45(a)–(b), the synchronization states on the x_1-x plane and x_2-y plane are shown for Q = 16 μs and T = 32 μs, yielding ratio Q/T = 50%.

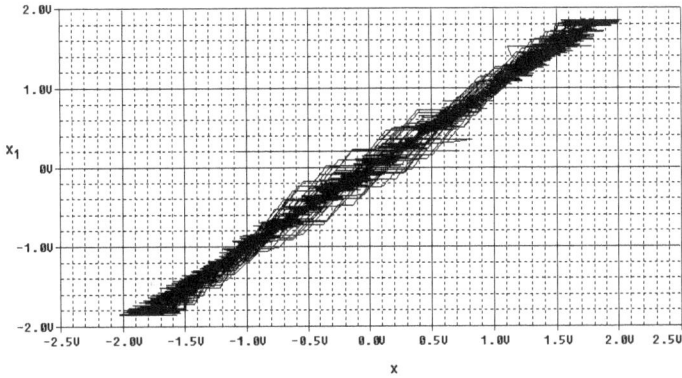

(a)

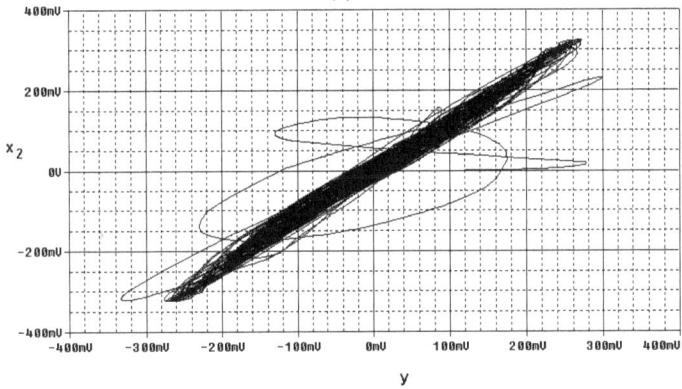

(b)

Fig. 6.45   For Q = 16 μs, T = 32 μs (Q/T = 50%), the synchronization graphs of the impulsive synchronization scheme in Fig. 6.42, (a) projection in x_1-x plane, (b) projection in x_2-y.

III. For T ≥ 40 μs

In this case, to achieve synchronization impulsively, the impulse width occupies at least 50% of the impulse period. In Fig. 6.46(a)–(b), the synchronization states on the x_1-x plane and x_2-y plane are shown for Q = 40 μs and T = 50 μs. yielding ratio Q/T = 80%.

(a)

(b)

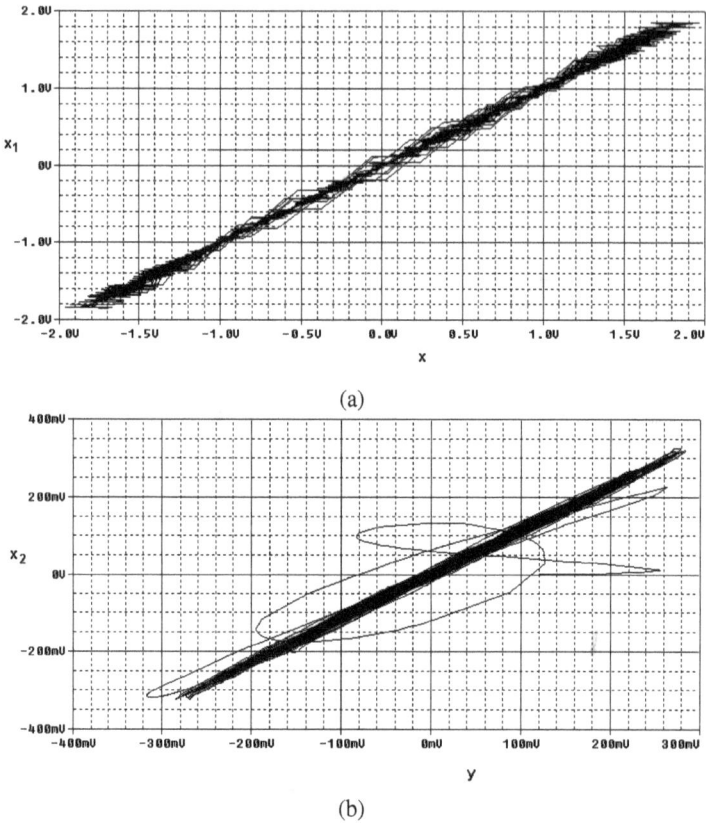

Fig. 6.46   For Q = 40 μs, T = 50 μs (Q/T = 80%), the synchronization graphs of the impulsive synchronization scheme in Fig.6.42, (a) projection in x_1-x plane, (b) projection in x_2-y plane.

6.5.4 *Experimental scheme for impulsive synchronization of two MMCCs*

The experimental setup of impulsive synchronization [64, 66] between two MMCCs is shown in Fig. 6.47. As shown in the figure, this impulsive synchronization scheme uses a single synchronizing impulse sequence, and impulsive synchronization is realized via V_{C1}-drive configuration. To synchronize two MMCCs impulsively, impulses sampled from chaotic dynamic of the driving circuit to the driven circuit are transmitted through one communication channel. When the switching signal s(t) is at a high level, the S5-switch is "ON" and a synchronizing impulse is transmitted from the driving system to the driven system. When the switching signal s(t) is at a low level, the switch is turned off, and the two MMCCs are disconnected.

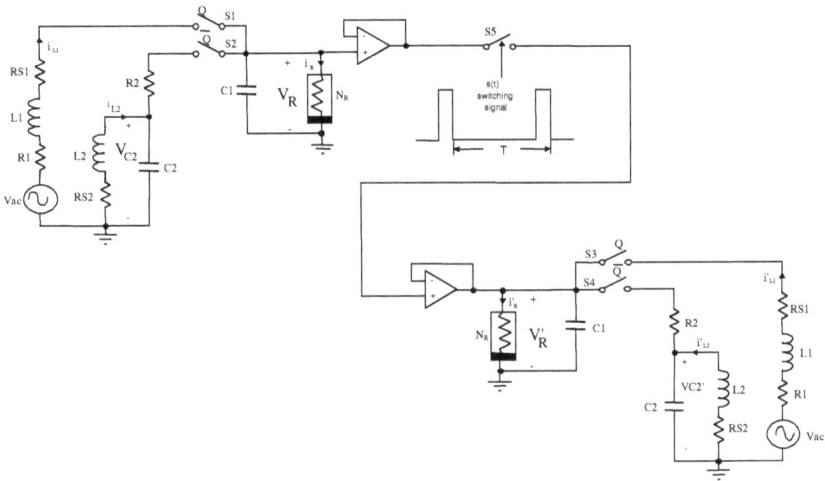

Fig. 6.47 Experimental setup of impulsive synchronization between two mixed-mode chaotic circuits.

6.5.4.1 *Experimental results*

Before the impulsive synchronization experiments with two MMCCs, we experimentally tested the system in Fig. 6.47 for full coupling, *i.e.*, continuous synchronization. The chaotic dynamics of driving and driven circuits and synchronization graphs obtained from laboratory experiments for autonomous mode (S1 and S3-OFF, S2 and S4-ON) and nonautonomous mode (S1 and S3-ON, S2 and S4-OFF) of each MMCC are given in Fig. 6.48 and Fig. 6.49, respectively. The experimental results obtained in dynamic operation of the system are illustrated in Fig. 6.50.

(a)

(b)

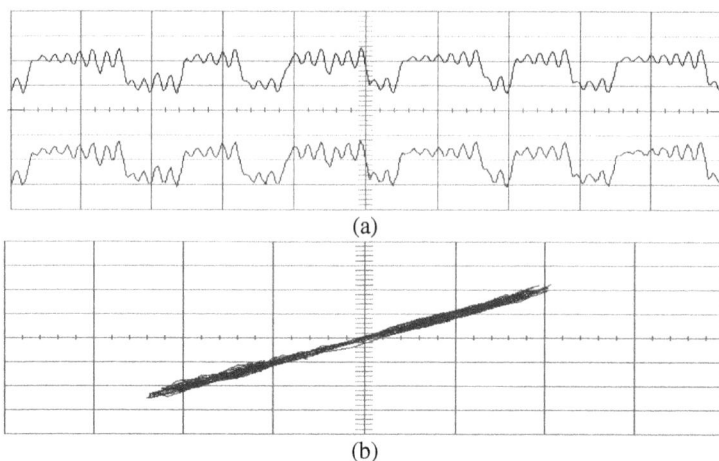

Fig. 6.48   For full coupling, in case of MMCCs operating in autonomous mode, experimental results of synchronization scheme in Fig. 6.47, (a) the upper trace V_{C1} (5V/div), the lower trace V_{C1}' (5 V/div) time/div: 2 ms/div, (b) synchronization graph observed in the V_{C1}- V_{C1}' plane for autonomous mode, x-axes: 2 V, y-axes: 2 V.

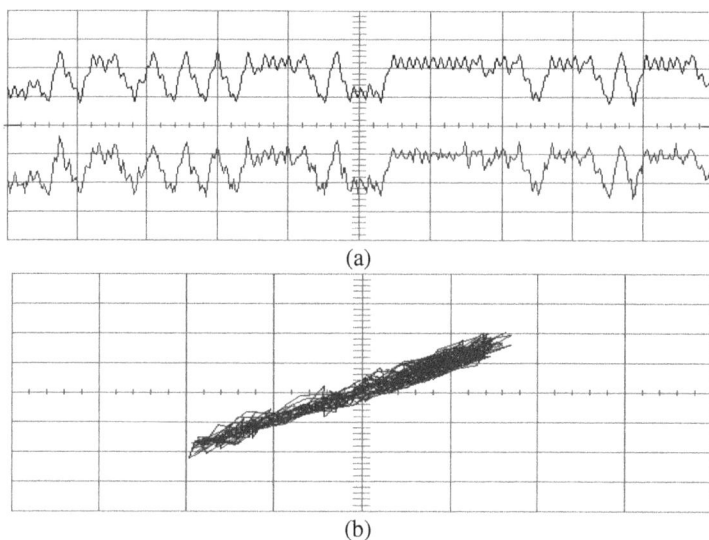

(a)

(b)

Fig. 6.49   For full coupling, in case of MMCCs operating in nonautonomous mode, experimental results of synchronization scheme in Fig. 6.47, (a) the upper trace V_{C1} (2 V/div), the lower trace V_{C1}' (2 V/div) time/div: 1 ms/div, (b) synchronization graph observed in the V_{C1}- V_{C1}' plane for autonomous mode, x-axes: 1 V, y-axes: 1 V.

(a)

(b)

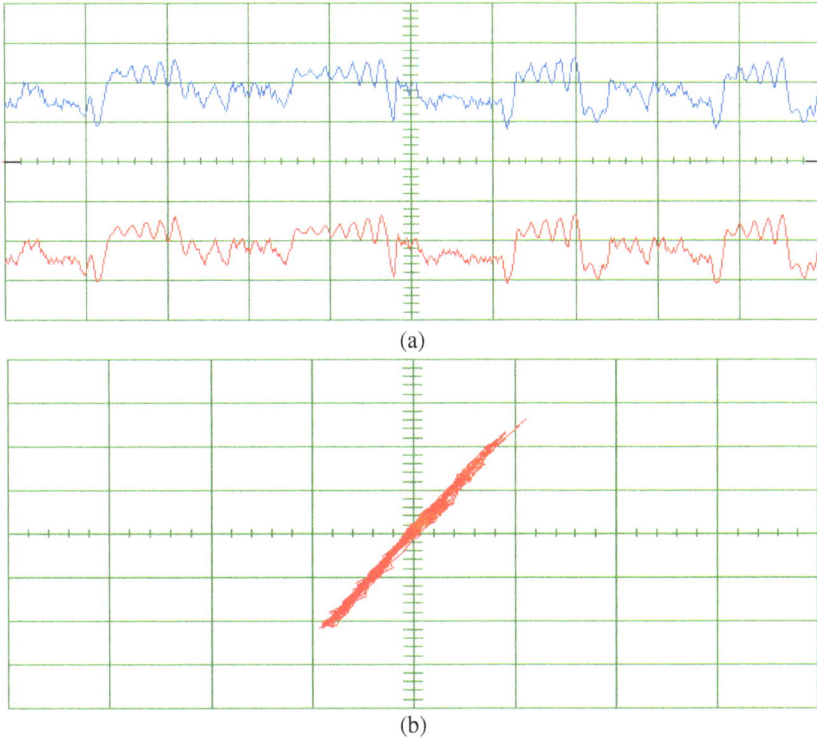

Fig. 6.50 For full coupling, in case of MMCCs operating in mixed-mode, experimental results of synchronization scheme in Fig. 1, (a) the upper trace V_{C1} (5V/div), the lower trace V_{C1}' (5 V/div) time/div: 2 ms/div, (b) synchronization graph observed in the V_{C1}-V_{C1}' plane, x-axes: 5 V, y-axes: 2 V.

These experimental results show that the circuits are well-synchronized in full coupling operation mode. After confirming the continuous synchronization of two MMCCs, we started our impulsive synchronization experiments by applying the switching signal at frequencies under 1 kHz. For the switching process, we used a voltage-controlled electronic switching device, namely, 4016 IC. For low-frequency impulsive progress of the system in Fig. 6.47, we observed that two circuits behave independently and consequently the synchronization does not occur. The experimental results for switching frequency f = 1 kHz are shown in Fig. 6.51.

(a)

(b)

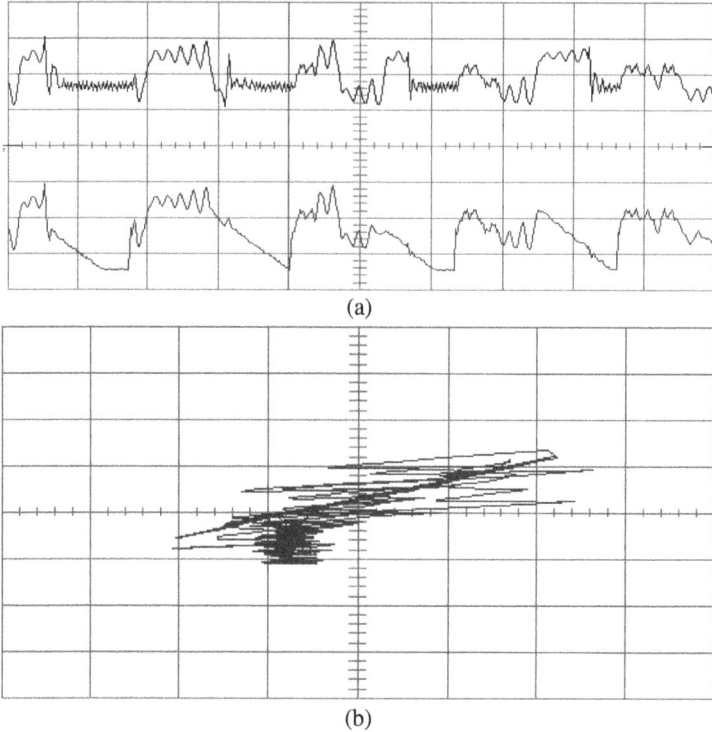

Fig. 6.51   Experimental results of impulsive synchronization scheme for switching frequency f = 1 kHz, (a) the upper trace V_{C1} (5 V/div), the lower trace $V_{C1}{'}$ (5 V/div) time/div: 2 ms/div, (b) synchronization graph observed in the V_{C1}- $V_{C1}{'}$ plane, x-axes: 2 V, y-axes: 5 V.

By increasing the switching frequency beyond f = 10 kHz, two MMCCs start to behave similarly, and the synchronization is achieved in robust form. Fig. 6.52 illustrates the experimental results for switching frequency f = 20 kHz. We have investigated the system with the switching signal up to 1 MHz. In these switching frequencies, two circuits exhibit nearly identical mixed-mode chaotic dynamics, and the synchronization is almost fully achieved as shown in the experimental results illustrated in Fig. 6.53 (for f = 100 kHz) and Fig. 6.54 (for f = 500 kHz).

(a)

(b)

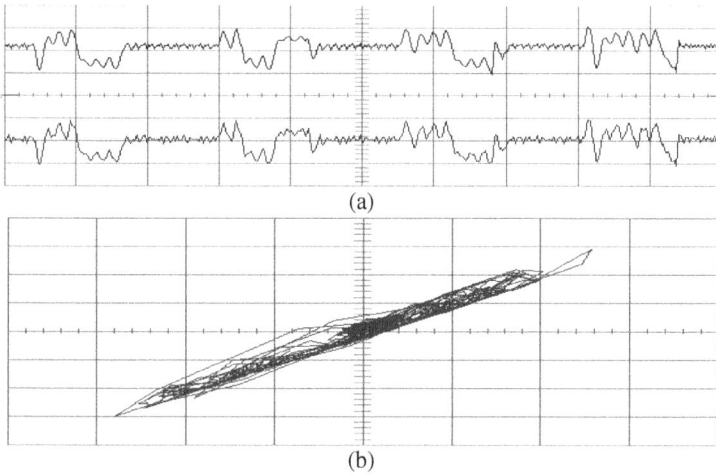

Fig. 6.52   Experimental results of impulsive synchronization scheme for switching frequency f = 20 kHz, (a) the upper trace V_{C1} (5 V/div), the lower trace V_{C1}' (5 V/div) time/div: 2 ms/div, (b) synchronization graph observed in the V_{C1}- V_{C1}' plane, x-axes: 2 V, y-axes: 2 V.

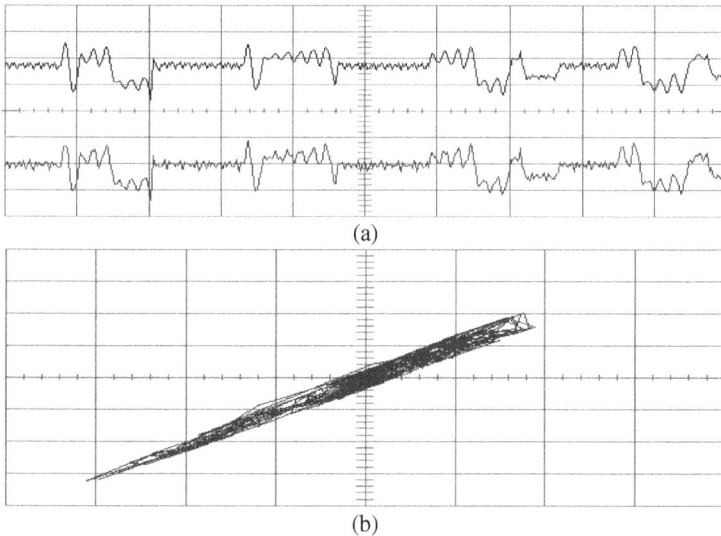

(a)

(b)

Fig. 6.53   Experimental results of impulsive synchronization scheme for switching frequency f = 100 kHz, (a) the upper trace V_{C1} (5 V/div), the lower trace, V_{C1}' (5 V/div) time/div: 2 ms/div, (b) synchronization graph observed in the V_{C1}- V_{C1}' plane, x-axes: 2 V, y-axes: 2 V.

(a)

(b)

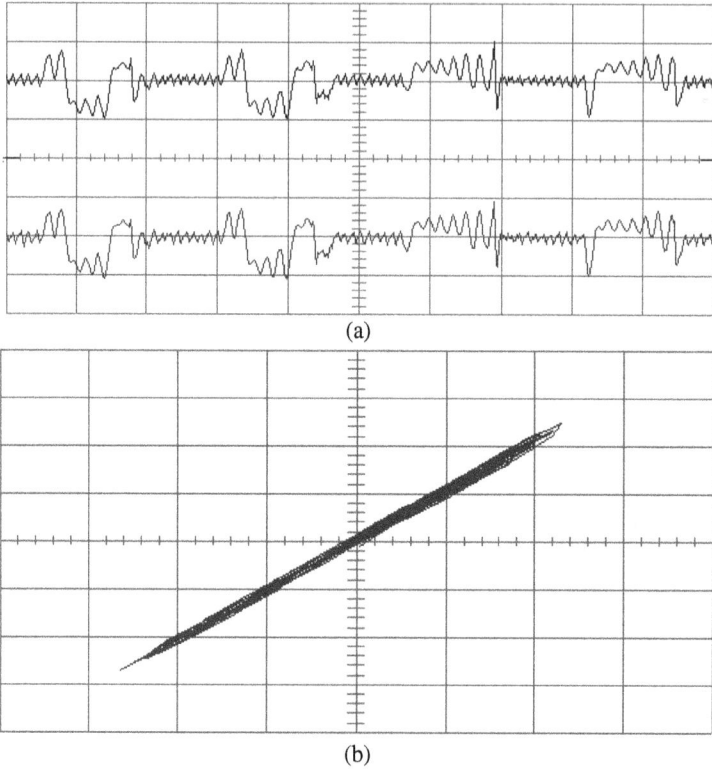

Fig. 6.54 Experimental results of impulsive synchronization scheme for switching frequency f = 500 kHz, (a) the upper trace V_{C1} (5 V/div), the lower trace, V_{C1} (5 V/div) time/div: 2 ms/div, (b) synchronization graph observed in the V_{C1}-V_{C1} plane, x-axes: 2 V, y-axes: 2 V.

Chapter 7

A Laboratory Tool for Studying Chua's Circuits

Although many authors have prepared useful papers to illustrate the existence of chaos and to enhance the reader understanding of chaos by using theoretical, simulation or experimental setups, no chaos-based laboratory work-board has been designed or implemented for studying chaotic circuits and systems. We designed and implemented a laboratory tool for studying Chua's circuits. In this chapter, we introduce this versatile laboratory tool.

7.1 Introduction

In the literature, many useful educational papers have been prepared by using theoretical, simulation and/or experimental setups to illustrate the existence of chaos and to enhance the reader understanding of chaos [3, 44, 90, 98, 137]. In these works, while simulation-based studies have been generally realized by circuit analysis simulation tools such as PSPICE, experimental studies include the laboratory reports on a specific chaotic circuit or a system. Although these theoretical and practical studies on chaotic circuits and systems are very helpful, laboratory tools designed in a systematic way are required for introducing chaotic circuits and systems in graduate and undergraduate research and education programs. No chaos work-board as a laboratory tool has been designed and implemented until now for this purpose. To meet this requirement, we designed and implemented a laboratory tool for studying chaotic circuits and systems [76–77]. The mixed-mode chaotic circuit [73] was a very suitable chaotic circuit model for this purpose. This is because autonomous, nonautonomous and mixed-mode chaotic dynamic types can all be investigated on this model. Also, alternative design ideas for Chua's diode and inductorless realization of Chua's diode-based chaotic circuits can be easily applied on this model.

7.2 Description of the Laboratory Tool

The photograph and layout of the implemented chaos training board are shown in Fig.7.1 and 7.2, respectively. This board is the laboratory trainer that enables students to investigate autonomous, nonautonomous and mixed-mode chaotic dynamics.

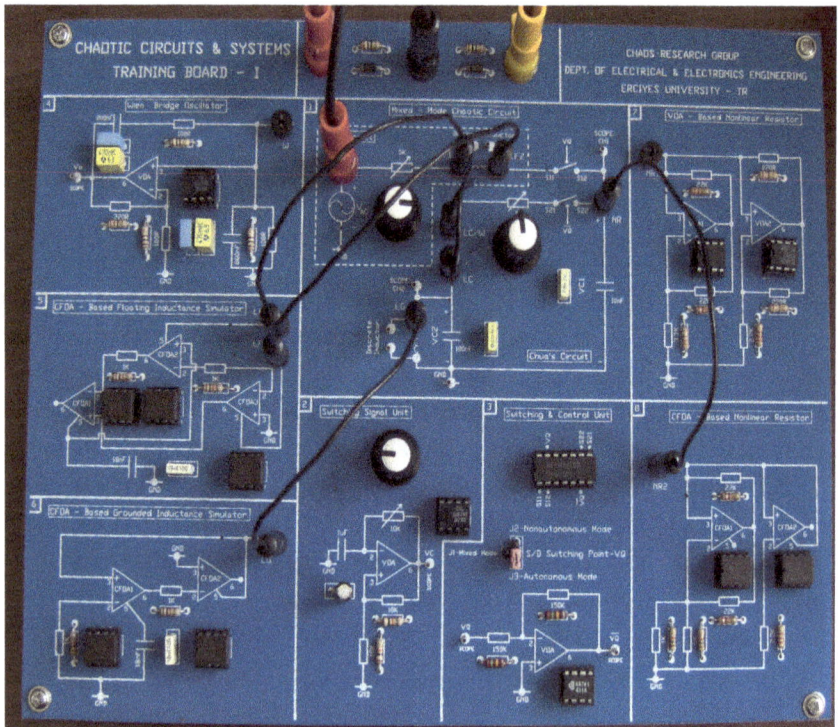

Fig. 7.1 Photograph of the nonlinear chaos training board.

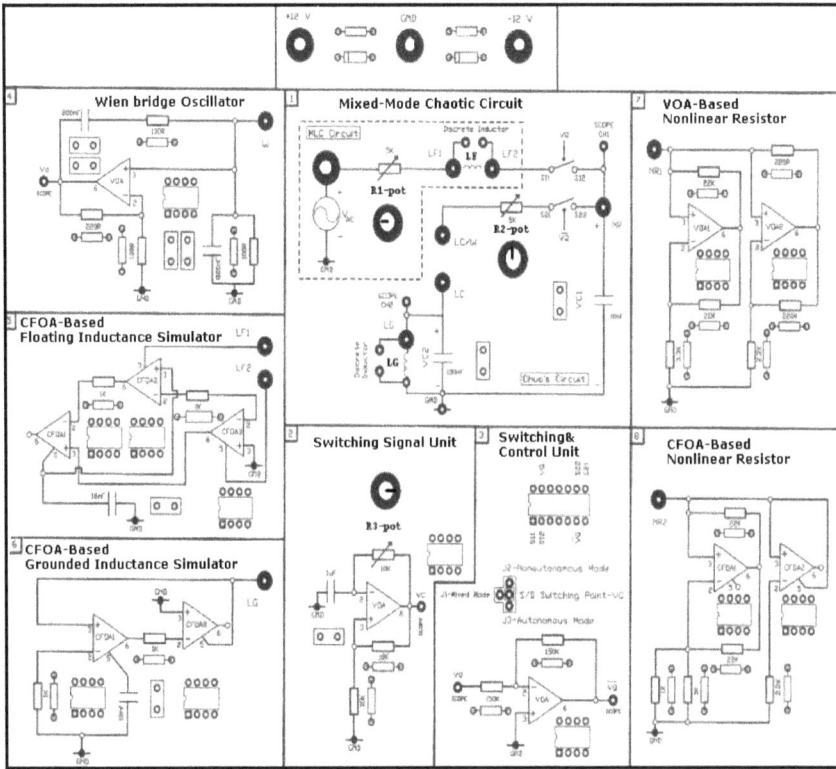

Fig. 7.2 The layout of the nonlinear chaos training board.

The chaos training board consists of eight preconstructed circuit blocks. The numbered blocks on the training board are labeled as follows:

❶ Mixed-Mode Chaotic Circuit (MMCC) Part
❷ Switching Signal Unit-A Square Wave Generator
❸ Switching & Control Unit
❹ Wien-Bridge Oscillator
❺ Current Feedback Operational Amplifier (CFOA)-Based Floating Inductance Simulator
❻ Current Feedback Operational Amplifier (CFOA)-Based Grounded Inductance Simulator

❼ Voltage Mode Operational Amplifier (VOA)-Based Nonlinear Resistor
❽ CFOA-Based Nonlinear Resistor

The board is energized by connecting an external DC symmetrical power supply to power sockets on top of the board. All of the ICs on the board are powered from these sockets. The core of the training board is the MMCC block shown in Fig. 7.3(a). A detailed analysis and description of the MMCC have been given in Chapter 4. As stated in Chapter 4, an MMCC operates either in the chaotic regime determined by the autonomous circuit part, or in the chaotic regime determined by the nonautonomous circuit part, depending on static switching, and it is capable of operating in mixed-mode, which includes both autonomous and nonautonomous regimes, depending on dynamic switching. Because of these versatile features, the MMCC offers an excellent educational circuit model for studying and practical experimenting on chaos and chaotic dynamics. The nonlinear resistor N_R is only a nonlinear element in MMCC and its piecewise i-v characteristic, shown graphically in Fig. 7.3(b), is defined by

$$i_R = f(V_R) = G_b V_R + 0.5 \cdot (G_a - G_b) \times (|V_R + B_p| - |V_R - B_p|) \quad (7.1)$$

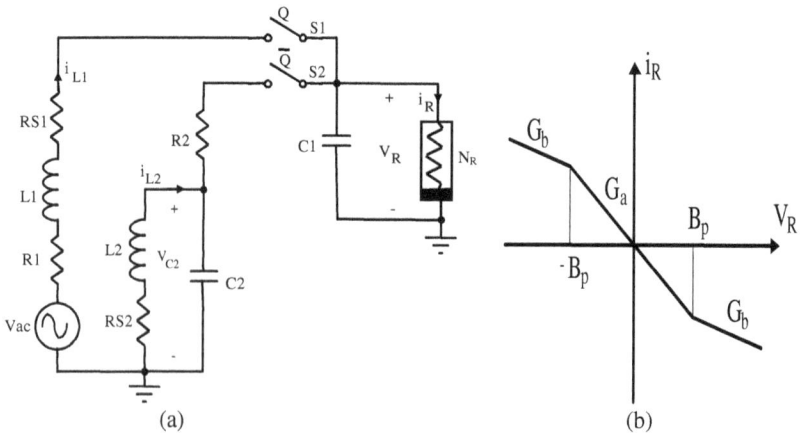

(a)   (b)

Fig. 7.3 (a) The circuit scheme of the MMCC, (b) i-v characteristic of nonlinear resistor in the MMCC.

In the literature, a variety of circuit topologies have been considered for realizing the nonlinear resistor. A good survey of the proposed circuit topologies for the nonlinear resistor, namely Chua's diode, has been given in Chapter 1. Of these realizations, a VOA-based nonlinear resistor structure, which is formed by connecting two voltage controlled negative impedance converters in parallel, and the CFOA-based nonlinear resistor structure shown in Fig. 7.4 have been located as preconstructed circuits to the right side of the board.

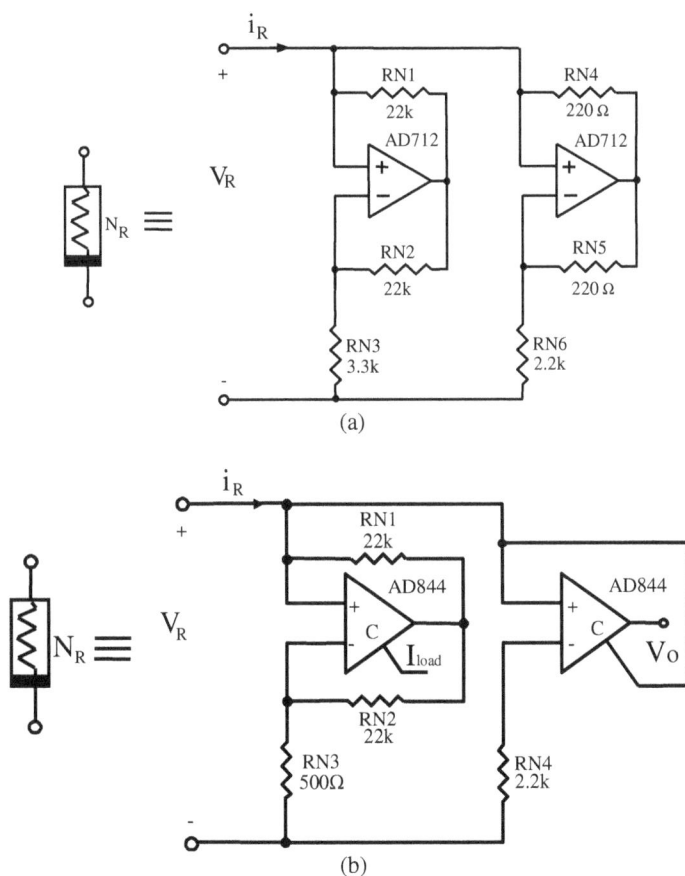

Fig. 7.4 (a) VOA-based nonlinear resistor, (b) CFOA-based nonlinear resistor.

In experimental studies with the proposed training board, the user can investigate the main chaotic circuit block by connecting either the VOA-based nonlinear resistor block (N_{R1}) or the CFOA-based nonlinear resistor block (N_{R2}) to the N_R-junction point on the training board as shown in Fig. 7.5. While the N_{R1} block is implemented by using two VOAs and six resistors, the alternative CFOA-based N_{R2} block is implemented by using two AD844 type CFOAs and four resistors.

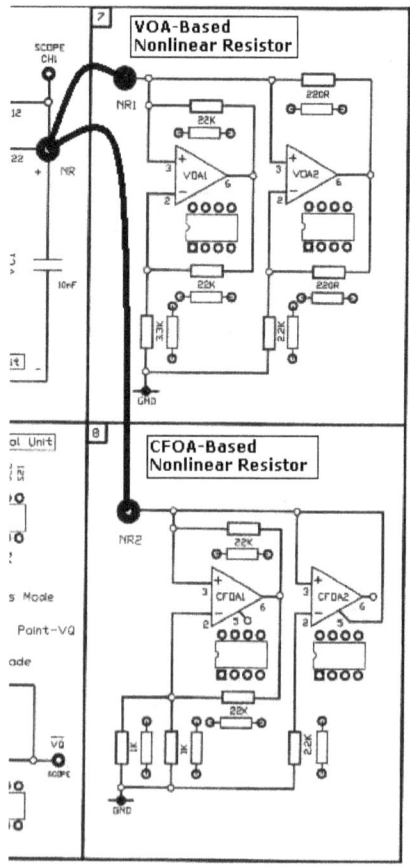

Fig. 7.5 Connection of two alternative nonlinear resistor blocks (N_{R1}, N_{R2}) to N_R junction point.

As a result of the flexible and versatile design methodology of the training board, the user can configure the main chaotic circuit block in two forms: First, the user can configure the circuit in the conventional manner by placing discrete inductor elements to related sockets on the training board. Alternatively, the user can configure the chaotic circuit in inductorless form by using CFOA-based grounded and floating inductance simulators and Wien-bridge oscillator blocks located to the left side of the board. The user can use the CFOA-based floating inductance simulator block for L_F floating inductor in MMCC by connecting the outputs of the simulator to L_{F1} and L_{F2} sockets. Similarly, the user can use the CFOA-based grounded inductance simulator block for L_G grounded inductor in MMCC by connecting the output of the simulator to L_G socket. In this process, an additional board connection between LC resonator part (LC-point) and LC/W junction point is required as shown in Fig. 7.6.

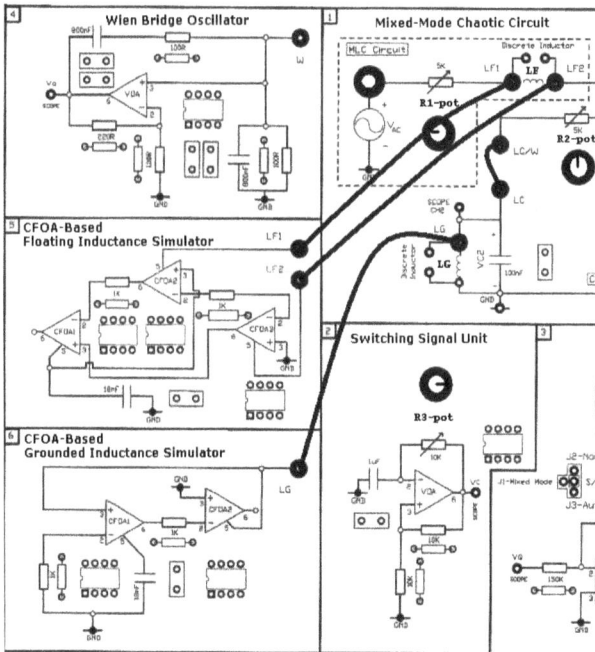

Fig. 7.6 Connection of CFOA-based floating and grounded inductance simulator blocks to floating and grounded inductor sockets on the board.

In this inductorless configuration, the user can also use the Wien bridge oscillator block instead of the LC resonator part on the training board. For this operation, the LC resonator part (LC-point) and LC/W junction point should be disconnected and the conjunction point (W-point) of the Wien bridge oscillator connected to the LC/W point in the MMCC block. This operation is illustrated in Fig. 7.7. In the training board, while the floating inductance simulator block is configured by three AD844 type CFOAs, two resistors R_{F1} and $R_{F2} = 1$ kΩ and a capacitor $C_F = 18$ nF to simulate $L_F = 18$ mH, the grounded inductance simulator block is configured by two AD844 type CFOAs, two resistors R_{G1} and $R_{G2} = 1$ kΩ and a capacitor $C_G =18$ nF to simulate $L_G = 18$ mH. Routine analysis for these simulators yields equivalent inductance as

$$L_{eq} = R_{F1} \cdot R_{F2} \cdot C_F = R_{G1} \cdot R_{G2} \cdot C_G \tag{7.2}$$

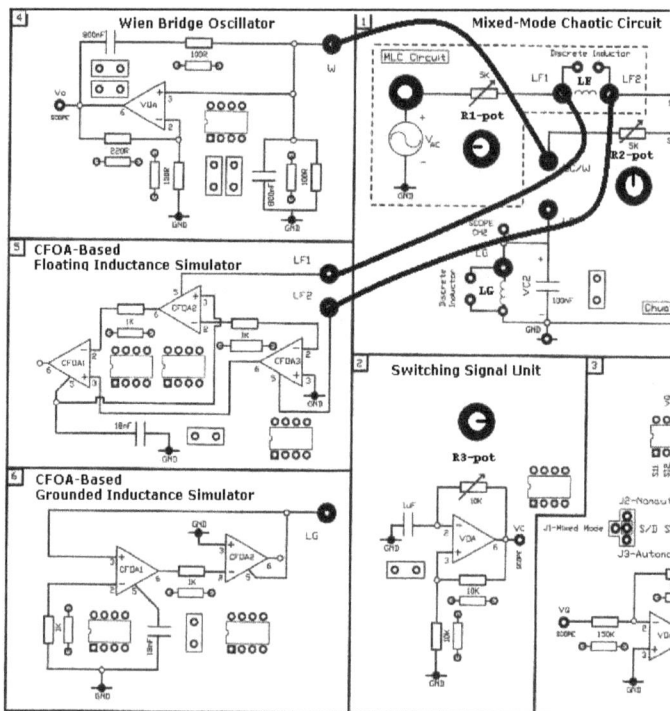

Fig. 7.7 Using Wien bridge oscillator block instead of LC resonator part in the MMCC.

The switching control of the MMCC is provided by the switching & control unit on the training board. A 4016 IC was used as the switching element. A common static and dynamic (S/D) switching point-V_Q can be connected to three positions via J1, J2 and J3 jumper adjustments on the board. These jumper adjustments are summarized in Table 7.1.

Table 7.1 Board configurations.

Jumper Adjustment on Board	Positions of Switches on Board	Chaotic Circuit Type Formed by Jumper Adjustment	Oscillation Mode
J3	Static Switching S1-OFF/S2-ON	Chua's Circuit	Autonomous
J2	Static Switching S1-ON/S2-OFF	MLC Circuit	Nonautonomous
J1	Dynamic Switching	MMCC	Mixed-Mode

When this point is connected to the J2 point via a jumper, V_Q control signal has a dc positive power supply level, and its complementary signal V'_Q has a dc negative power supply level via the VOA-based inverter circuit on the board. Hence in this static switching situation, the positions of S1 and S2 become ON and OFF respectively, yielding a nonautonomous operation mode of the MMCC. The user can investigate the nonautonomous chaotic dynamics of a MLC circuit in this operation mode. J3 jumper adjustment reverses this process. With the J3 jumper adjustment, the V_Q control signal has negative dc voltage, and its complementary signal V'_Q has positive dc voltage. Hence, in this static switching situation the positions of S1 and S2 become OFF and ON, respectively, yielding an autonomous operation mode of the MMCC. The user can investigate the autonomous chaotic dynamics of Chua's circuit in this operation mode. Besides static switching, there is also dynamic switching configuration via J1 jumper adjustment. With this J1 adjustment, S/D switching point-V_Q is connected to the output (V_C) of the square wave generator on the board. Hence, a dynamic switching operation is obtained yielding mixed-mode operation, and the positions of S1 and S2 switches change continuously according to the frequency of the square wave generator. As a result of this operation, the user will be introduced to a very interesting phenomenon, namely mixed-mode chaotic dynamics, which include both autonomous and nonautonomous chaotic dynamics.

7.3 Experimental Studies with the Work-Board

Researchers interested in nonlinear science and chaos can use this work-board for investigation of autonomous, nonautonomous and mixed-mode chaotic dynamics. The configuration of the proposed work-board is very simple and versatile. By applying the jumper adjustments defined in Table 7.1, the users can easily realize laboratory experiments on relevant chaotic circuits. The switching control of the MMCC circuit is provided by the switching & control unit, in which a 4016 IC is used as the switching element.

In this section, we will give some sample experiments and experimental results obtained from the chaos work-board configured in autonomous, nonautonomous and mixed-mode forms via the jumper adjustments in Table 1. For monitoring V_{C1} and V_{C2} chaotic dynamics at an oscilloscope screen, CH1 and CH2 pins were located on the work-board.

7.3.1 *Experimental measurement of v-i characteristics of VOA-based and CFOA-based nonlinear resistors on the training board*

The v-i characteristics of two VOA-based and CFOA-based Chua's diodes on the board can be measured by constructing an additional experimental setup shown in Fig. 7.8. In the described configuration, V_{tr} is a triangular waveform with zero dc offset, amplitude 7 V peak-to-peak, and frequency 30 Hz. As shown in the figure, a current-sensing resistor-R_{sense} is used to measure the current I_R which flows into the VOA-based or CFOA-based Chua's diode. The value of R_{sense} can be determined as 100Ω–500Ω. By applying a V_{tr} voltage source, the v-i characteristic of Chua's diode can be measured. For this measurement process, V_{IR} is connected to the Y-input and V_R is connected to the X-input of an oscilloscope in X-Y mode. Note that in this measurement, the current I_R is measured indirectly. The measured dc characteristics of VOA-based and CFOA-based Chua's diodes on the board are shown in Fig. 7.9.

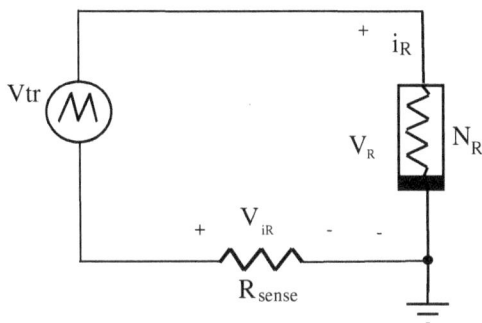

Fig. 7.8 Experimental setup for measuring the v-i characteristics of two Chua's diode configurations on the board.

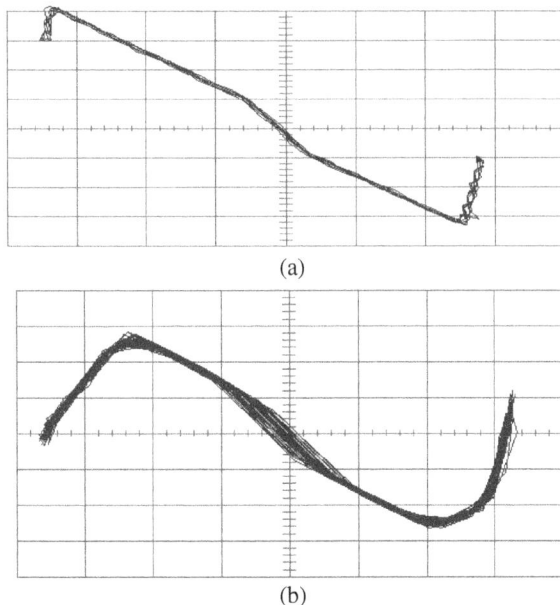

(a)

(b)

Fig. 7.9 The measured dc characteristics of (a) VOA-based Chua's diode; x-axes: 2 V/div, y-axes: 500 mV/div, (b) CFOA-based Chua's diode; x-axes: 2 V/div, y-axes: 500 mV/div.

7.3.2 *Investigation of autonomous chaotic dynamics via training board*

In order to investigate autonomous chaotic dynamics via the training board, the board connections can be done as shown in Fig. 7.10. For this operation mode, at first the S/D switching point V_Q is positioned to the J3 jumper point. As a nonlinear resistor, the CFOA-based nonlinear resistor block N_{R2} is connected to the N_R point in the MMCC. With this training board configuration, inductorless autonomous Chua's circuit configuration is obtained, and the user can investigate the autonomous chaotic dynamics on this configuration. With this configuration, the MMCC is governed by the Chua's circuit equations defined in Chapter 4.

By adjusting the R2 potentiometer in the MMCC, the user can observe a Hopf-like bifurcation from dc equilibrium, a sequence of period-doubling bifurcations to double-band Chua chaotic attractor and a boundary limit.

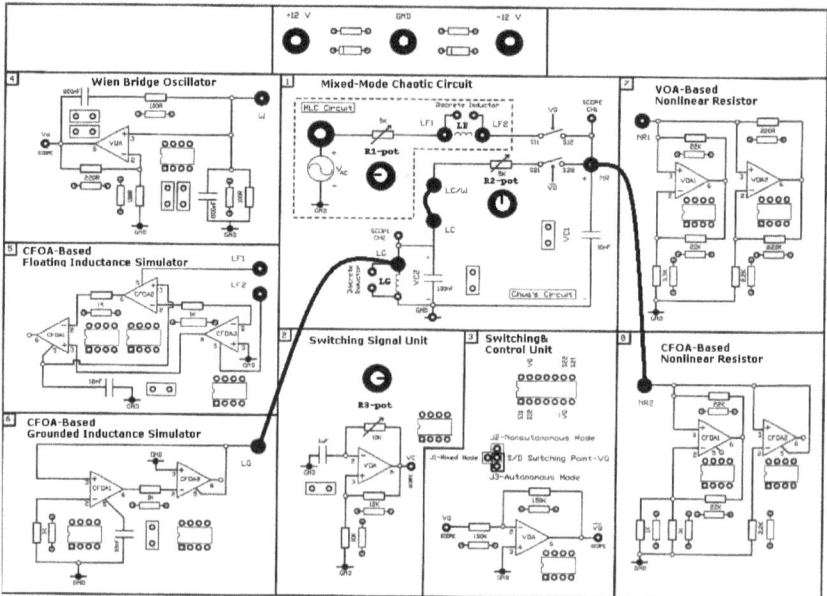

Fig. 7.10 Training board configuration for investigation of autonomous chaotic dynamics.

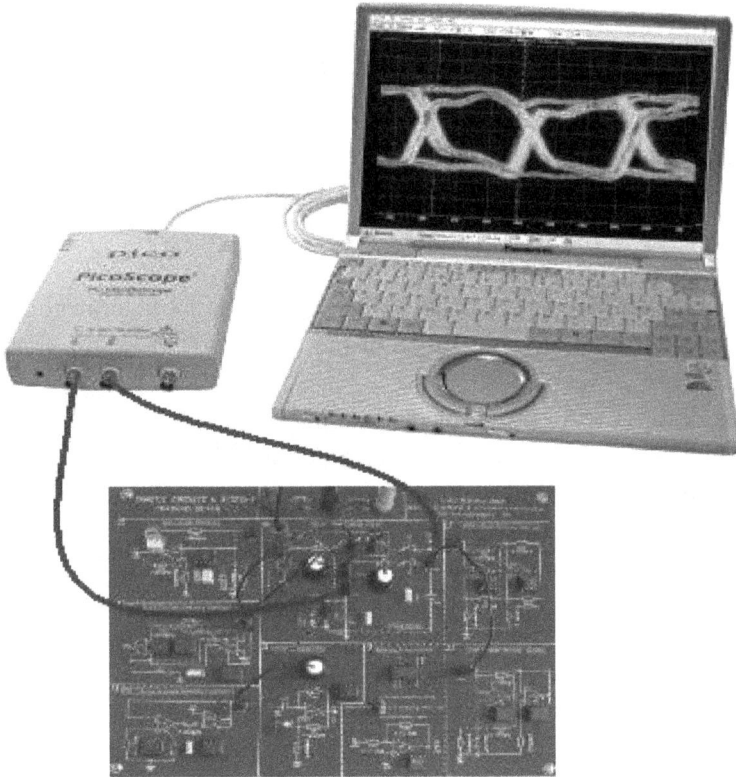

Fig. 7.11 A PC-compatible virtual measurement system using a PC oscilloscope module for the chaos training board.

In addition to use of a digital storage oscilloscope as in our laboratory experiments, a more versatile and useful laboratory measurement system as shown in Fig. 7.11 can be used to monitor chaotic behaviors from the board. This system is a PC-compatible virtual measurement system using a PC oscilloscope module [114]. The PC oscilloscope module incorporates software that turns into an oscilloscope and spectrum analyzer. This system is flexible, easy to use and has many advantages over conventional instruments, including multiple views of the same signal and on-screen display of voltage and time. The features of the PC oscilloscope module used in the virtual measurement system in our experiments are listed below:

Bandwidth	200 mHz
Sampling rate	10 GS/s
Channels	2 + Ext. trigger/signal gen.
Oscilloscope timebases	1 ns/div to 50 s/div
Spectrum ranges	0 to 100 mHz
Record length	1 MB
PC connection	USB 2.0 (USB1.1 compatible)

Some experimental observations via the virtual measurement system in Fig. 7.11 for autonomous mode of the board have been illustrated in Fig. 7.12.

The design methodology of the training board gives the user several alternative ways for configuring the autonomous mode. The user can realize the same experiments on an autonomous Chua's circuit by exchanging the CFOA-based nonlinear resistor with the VOA-based resistor. Similarly, the same experiments can be realized by using discrete inductor elements instead of inductance simulator blocks. Each of these alternative configurations can be arranged as separate laboratory experiments.

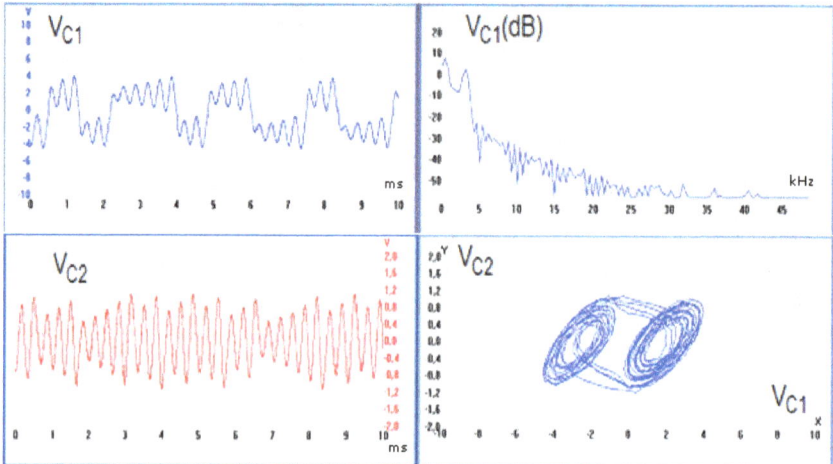

Fig. 7.12 Experimental measurements of MMCC on the board with the autonomous mode configuration in Fig. 7.10.

7.3.3 *Investigation of nonautonomous chaotic dynamics via training board*

To investigate nonautonomous chaotic dynamics via the training board, the board connections can be done as shown in Fig. 7.13. For this operation mode, at first the S/D switching point V_Q is positioned to the J2 jumper point on the board. With this board configuration, the nonautonomous MLC circuit configuration is obtained, and the user can investigate nonautonomous chaotic dynamics on this circuit configuration. V_{AC} sinusoidal signal is taken from sine-wave output of an external function generator. Its amplitude and frequency are determined as $V_{rms} = 100$ mV and $f = 8890$ Hz, respectively. With this configuration, the MMCC on the board is governed by the nonautonomous MLC circuit equations defined in Chapter 4. By adjusting the amplitude of the AC signal source and/or R_1 potentiometer located in the nonautonomous part of the MMCC, the user can easily observe the complex dynamics of bifurcation and the chaos phenomenon.

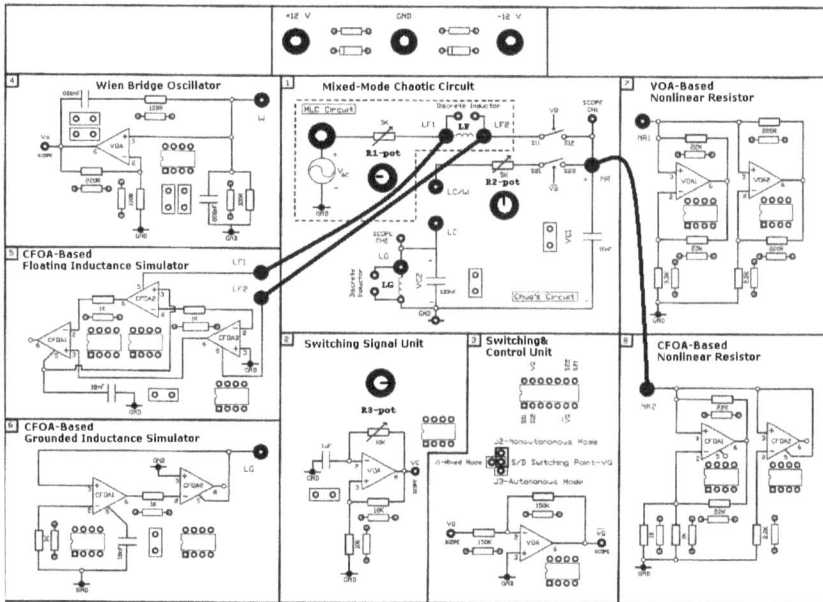

Fig. 7.13 Training board configuration for investigation of nonautonomous chaotic dynamics.

Some experimental observations via the virtual measurement system in Fig. 7.11 for this nonautonomous mode of the board have been illustrated in Fig. 7.14 through 7.17.

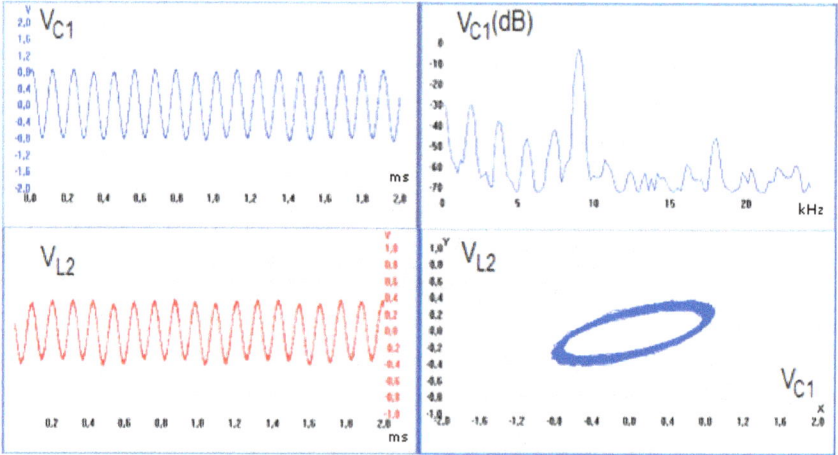

Fig. 7.14 Experimental measurements of MMCC on the board with the nonautonomous mode configuration in Fig. 7.13 (period-1).

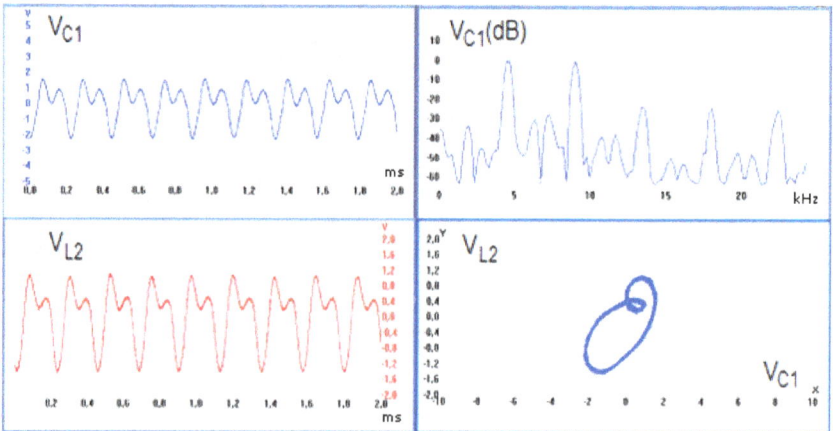

Fig. 7.15 Experimental measurements of MMCC on the board with the nonautonomous mode configuration in Fig. 7.13 (period-2).

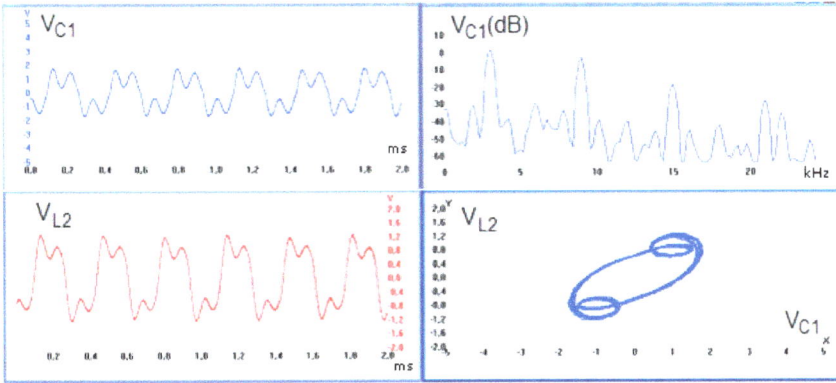

Fig. 7.16 Experimental measurements of MMCC on the board with the nonautonomous mode configuration in Fig. 7.13 (period-3).

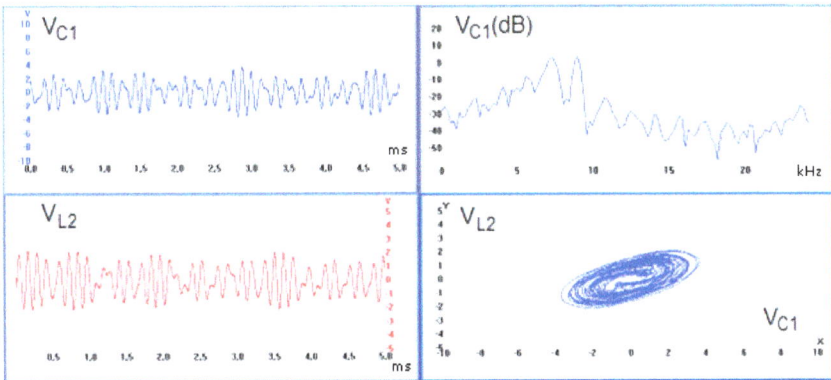

Fig. 7.17 Experimental measurements of MMCC on the board with the nonautonomous mode configuration in Fig. 7.13. (chaotic behavior)

As in the autonomous mode configuration of the board, the user can realize the same experiments by using the VOA-based nonlinear resistor block instead of the CFOA-based nonlinear resistor block, and/or the user can establish the experiments with a discrete inductor element instead of the floating inductance simulator. These arrangements can also be easily adopted as new laboratory assignments.

7.3.4 *Investigation of mixed-mode chaotic dynamics via training board*

To investigate mixed-mode chaotic dynamics via the training board, the board connections can be done as shown in Fig. 7.18. For this mixed-mode operation, the S/D switching point V_Q is connected to the output of a square-wave generator via J1-jumper adjustment. Due to this square-wave control, dynamic and continuous switching operation is provided, and the switching time which determines the durations of autonomous and nonautonomous chaotic oscillations can be easily adjusted via the R3 potentiometer located in the square-wave generator block.

In this mode, the user will be able to observe a very interesting phenomenon, namely the mixed-mode chaotic phenomenon, which includes both autonomous and nonautonomous chaotic dynamics.

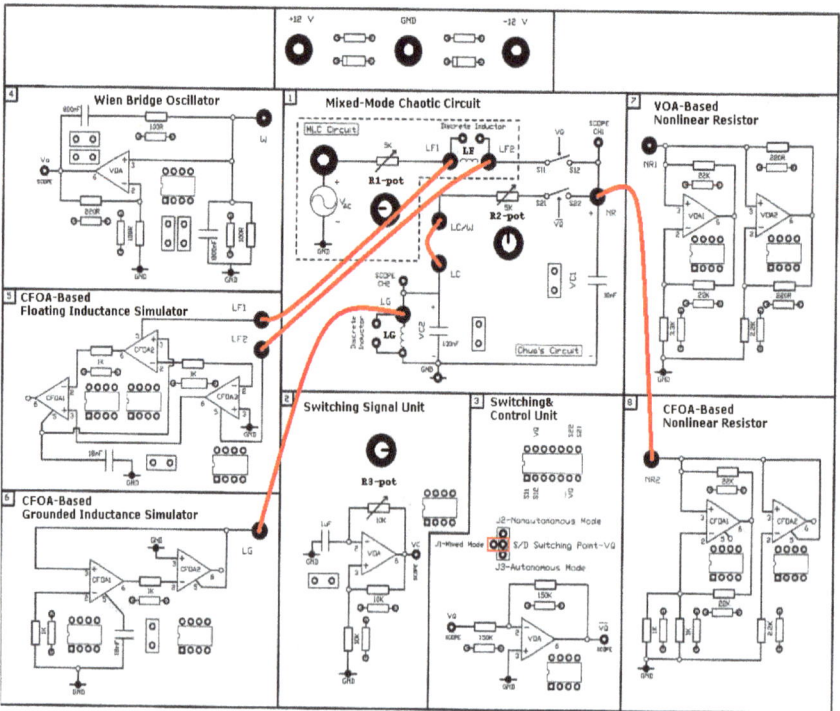

Fig. 7.18　Training board configuration for investigation of mixed-mode chaotic dynamics.

By adjustments of R_1 and R_2 potentiometers in the MMCC, a variety of mixed-mode chaotic dynamics are observed. Some experimental observations via the virtual measurement system in Fig. 7.11 for this mixed mode of the board are shown in Fig. 7.19 and 7.20. The other alternative configurations referred to in earlier assignments for nonlinear resistor and inductive structure can also be used as new experiments.

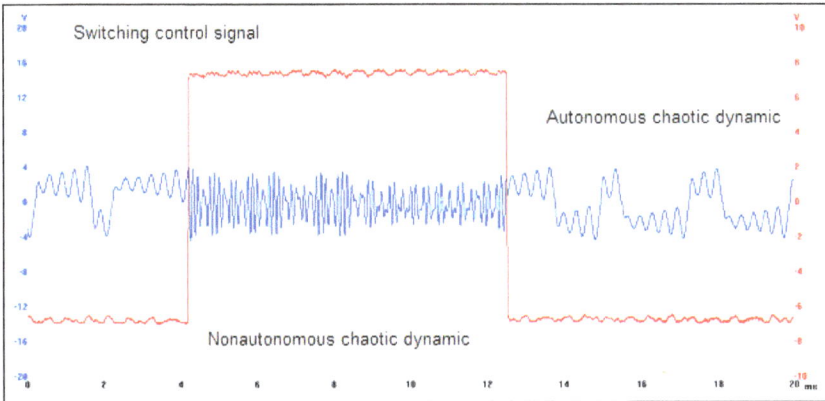

Fig. 7.19 Experimental measurements of MMCC on the board with the mixed-mode configuration in Fig. 7.18.

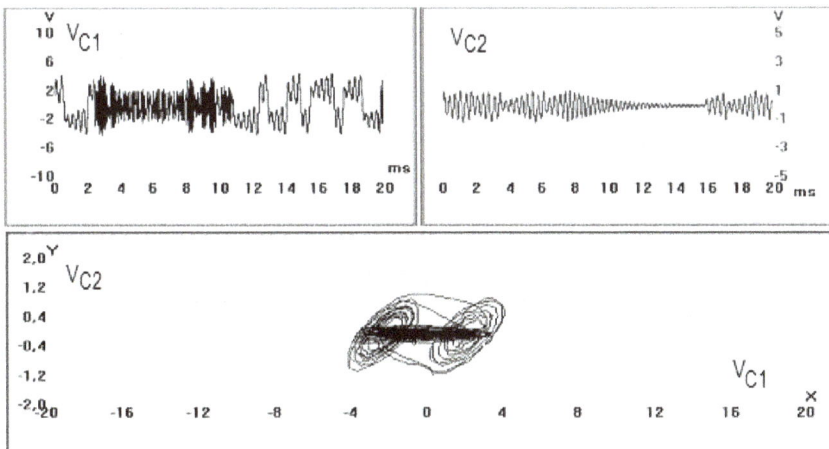

Fig. 7.20 Experimental measurements of MMCC on the board with the mixed-mode configuration in Fig. 7.18.

Bibliography

[1]   M. Ababneh, J-G. Barajas-Ramírez, G. Chen and L.S. Shieh. Robust digital controllers for uncertain chaotic systems: A digital redesign approach. *Chaos, Solitons & Fractals*, 31(5), pp. 1149-1164, 2007.

[2]   M.T. Abuel'matti and M.K. Aiyad. Chaos in an autonomous active-R circuit. *IEEE Trans. on Circuits & Syst.-I*, 42(1), pp. 1-5, 1995.

[3]   C. Aissi. Introducing chaotic circuits in an undergraduate electronic course. *Proc. of the 2002 ASEE Gulf-Southwest Annual Conference*, Lafayette, USA, 2002.

[4]   A.A. Alexeyev, G.V. Osipov and V.D. Shalfeev. Effects of square-wave modulation on CNN patterns. *IEEE Trans. on Circuits & Syst.-I*, 42(10), pp. 700-705, 1995.

[5]   Anadigm:*www.anadigm.com*

[6]   F.T. Arecchi, L. Fortuna and M. Frasca. A programmable electronic circuit for modeling CO_2 laser dynamics. *Chaos*, 15(043104), 2005.

[7]   P. Arena, S. Baglio, L. Fortuna and G. Manganaro. Chua's circuit can be generated by CNN cells. *IEEE Trans. on Circuits & Syst.-I*, 42(2), pp. 123-125, 1995.

[8]   P. Arena, A. Buscarino, L. Fortuna and M. Frasca. Separation and sysnchronization of piecewise linear chaotic systems. *Phys. Rev.*, E74(026212), 2006.

[9]   A. Arulgnanam, K. Thamilmaran and M. Daniel. Chaotic dynamics with high complexity in a simplified new nonautonomous nonlinear electronic circuit. *Chaos, Solitons & Fractals*, 42(4), pp. 2246-2253, 2009.

[10]   I.A. Awad and A.M. Soliman. Inverting second generation current conveyors: the missing building blocks,CMOS realizations and applications. *Int. Journal of Electronics*, 86(4), pp. 413-432, 1999.

[11]   E-W. Bai, K.E. Lonngren and J.C. Sprott. On the synchronization of a class of electronic circuits that exhibit chaos. *Chaos, Solitons & Fractals*, 13(7), pp. 1515-1521, 2002.

[12]   A. Buscarino, L. Fortuna and M. Frasca. Experimental separation of chaotic signals through synchronization. *Phil. Trans. R. Soc. A.*, 366, pp. 569-577, 2008.

[13]   S. Callegari, R. Rovatti and G. Setti. First direct implementation of a true random source on programmable hardware. *Int. Journal of Circ. Theor. Appl.*, 33, pp. 1-16, 2005.

[14] U. Çam and H. Kuntman. A new CMOS realisation of four terminal floating nullor (FTFN). *Int. Journal of Electronics*, 87(7), pp. 809-817, 2000.

[15] R. Caponetto, A.D. Mauro, L. Fortuna and M.Frasca. Field Programmable analog array to implement a programmable Chua's circuit. *Int. Journal of Bifurcation and Chaos*, 15, pp. 1829-1836, 2005.

[16] R. Caponetto, M. Lavorgna and L. Occhipinti. Cellular neural networks in secure transmission applications. *Proc. of CNNA'96*, pp. 411-416, 24-26 June, Sevilla, Spain, 1996.

[17] R. Caponetto, M. Criscione, L. Fortuna, D. Occhipinti and L. Occhipinti. Programmable chaos generator, based on CNN architectures, with applications in chaotic communications. *Proc. of CNNA'98*, pp. 124-129, 14-17 April, London, England, 1998.

[18] J.C.D. Cardoso, H.A. Albuquerque, R.M. Rubinger. Complex periodic structures in bi-dimensional bifurcation diagrams of a RLC circuit model with a nonlinear NDC device. *Physics Letters A*, 373(23-24), pp. 2050-2053, 2009.

[19] G. Chen and T. Ueta, (Eds.), Chaos in Circuits and Systems, *World Scientific*, 2002.

[20] T-I. Chien and T-L. Liao. Design of secure digital communication systems using chaotic modulation, cryptography and chaotic synchronization. *Chaos, Solitons & Fractals*, 24(1), pp. 241-255, 2005.

[21] L.O. Chua, CNN: A Paradigm for complexity. *World Scientific*, 1998.

[22] L.O. Chua, C.A. Desoer & E.S. Kuh. Linear and nonlinear circuits. *McGraw-Hill*, 1987.

[23] L.O. Chua, M. Itoh, L. Kocarev and K. Eckert. Chaos synchronization in Chua's circuit. *Journal of Circuits, Systems and Computers*, 3(1), pp. 93-108, 1993.

[24] L.O. Chua, C.W. Wu, A. Huang and G.A. Zhong. Universal circuit for studying and generating chaos. *IEEE Trans. on Circuits & Syst.*, CAS-40(10), pp. 732-745, 1993.

[25] L.O. Chua and L. Yang. Cellular neural networks: Theory. *IEEE Trans. on Circuits & Syst.*, 35, pp. 732-745, 1988.

[26] L.O. Chua, T. Yang, G-Q. Zhong and C-W. Wu. Adaptive synchronization of Chua's oscillators. *Int. Journal of Bifurcation & Chaos*. 6(1), pp. 189-201, 1996.

[27] L.O. Chua, T. Yang, G-Q. Zhong and C-W. Wu. Synchronization of Chua's circuits with time-varying channels and parameters. *IEEE Trans. on Circuits & Syst.-I*, 43(10), pp. 862-868, 1996.

[28] N.J. Corron and D.W. Hahs. A new approach to communications using chaotic signals. *IEEE Trans. on Circuits & Syst.-I*, 44(5), pp. 373-382, 1997.

[29] J.M. Cruz and L.O. Chua. A CMOS IC nonlinear resistor for Chua's circuit. *IEEE Trans. on Circuits & Syst.-I*, 39(12), pp. 985-995, 1992.

[30] K.M. Cuomo and A.V. Oppenheim. Circuit implementation of synchronized chaos with applications to communication. *Phys. Rev. Lett.*, 71(1), pp. 65-68, 1993.

[31] H. Dedieu, M.P. Kennedy and M. Hasler. Chaos shift keying: Modulation and demodulation of a chaotic carrier using self-synchronizing Chua's circuits. *IEEE Trans. on Circuits & Syst.-II*, 40(10), pp. 634-642, 1993.

[32] M. Drutarovsky and P. Galajda. Chaos-based true random number generator embedded in a mixed-signal reconfigurable hardware. *Journal of Electrical Engineering*, 57(4), pp. 218-225, 2006.

[33] A.S. Elwakil and M.P. Kennedy. Improved implementation of Chua's chaotic oscillator using current feedback op amp. *IEEE Trans. on Circuits & Syst.-I*, 47(1), pp. 76-79, 2000.

[34] A.S. Elwakil and M.P. Kennedy. Construction of classes of circuit-independent chaotic oscillators using passive-only nonlinear devices. *IEEE Trans. on Circuits & Syst.-I*, 48(3), pp. 289-306, 2001.

[35] A.S. Elwakil and A.M. Soliman. A family of Wien-type chaotic oscillators modified for chaos. *Int. Journal Circuit Theory and Appl.*, 25, pp. 561-579, 1997.

[36] T. Floyd, Electronic devices, *Pearson Prentice Hall*, 2005.

[37] P. Galajda and D. Kocur. Chua's circuit in spread spectrum communication systems. *Radioengineering*, 11(2), pp. 6-10, 2002.

[38] L. Gámez-Guzmán, C. Cruz-Hernández, R.M. López-Gutiérrez and E.E. García-Guerrero. Synchronization of Chua's circuits with multi-scroll attractors: Application to communication, *Communications in Nonlinear Science and Numerical Simulation*, 148(6), pp. 2765-2775, 2009.

[39] T. Gao, Z. Chen, Q. Gu and Z. Yuan. A new hyper-chaos generated from generalized Lorenz system via nonlinear feedback. *Chaos, Solitons & Fractals*, 35(2), pp. 390-397, 2008.

[40] J.M. Gonzalez-Miranda. Synchronization and control of chaos. *Imperial College Press*, 2004.

[41] L. Goras and L.O. Chua. Turing patterns in CNNs-II: Equations and behaviors. *IEEE Trans. on Circuits & Syst.-I*, 42(10), pp. 612-626, 1995.

[42] C. Güzeliş and L.O. Chua. Stability analysis of generalized cellular neural networks. *Int. Journal Circuit Theory and Appl.*, 21, 1-33, 1993.

[43] K.S. Halle, C-W. Wu, M. Itoh and L.O. Chua. Spread spectrum communication through modulation of chaos. *Int. Journal of Bifurcation and Chaos*, 3(2), pp. 469-477, 1993.

[44] D.C. Hamill. Learning about chaotic circuits with SPICE. *IEEE Trans. on Education*, 36(1), pp. 28-35, 1993.

[45] M.P. Hanias and G.S. Tombras. Time series cross prediction in a single transistor chaotic circuit. *Chaos, Solitons & Fractals*, 41(3), pp. 1167-1173, 2009.

[46] A.S. Hegazi and H.N. Agiza and M.M. El-Dessoky. Adaptive synchronization for Rossler and Chua's circuit systems. *Int. Journal of Bifurcation and Chaos*, 12(7), pp. 1579-1597, 2002.

[47] M. Higashimura. Realisation of current-mode transfer function using four-terminal floating nullor. *Electronic Letters*, 27(2), pp. 170-171, 1991.

[48] J. Hu, S. Chen and L. Chen. Adaptive control for anti-synchronization of Chua's chaotic system. *Physics Letters A*, 339(6), pp. 455-460, 2005.

[49] G. Hu, L. Pivka and A.L. Zheleznyak. Synchronization of a one-dimensional array of Chua's circuits by feedback control and noise. *IEEE Trans. on Circuits & Syst.-I*, 42(10), pp. 736-740, 1995.

[50] C-K. Huang, S-C. Tsay and Y-R. Wu. Implementation of chaotic secure communication systems based on OPA circuits. *Chaos, Solitons & Fractals*, 23(2), pp. 589-600, 2005.

[51] M. Hulub, M. Frasca, L. Fortuna and P. Arena. Implementation and synchronization of 3×3 grid scroll chaotic circuits with analog programmable devices. *Chaos*, 16(013121), 2006.

[52] T. Irita, T. Tsujita, M. Fujishima and K. Hoh. A simple chaos-generator for neuron element utilizing capacitance–*npn*-transistor pair. *Computers and Electrical Engineering*, 24(1-2), pp. 43-61, 1998.

[53] M. Itoh. Synthesis of electronic circuits for simulating nonlinear dynamics. *Int. Journal of Bifurcation and Chaos*, 11(3), pp. 605-653, 2001.

[54] M. Itoh, H. Murakami and L.O. Chua. Communication systems via chaotic modulations. *IEICE Trans. Fund.*, E77-A(6), pp. 1000-1006, 1994.

[55] M. Itoh, T. Yang and L.O. Chua. Experimental study of impulsive synchronization of chaotic and hyperchaotic circuits. *Int. Journal of Bifurcation and Chaos*, 9(7), pp. 1393-1424, 1999.

[56] N. Jiang, W. Pan, L. Yan, B. Luo, L. Yang, S. Xiang and D. Zheng, Two chaos synchronization schemes and public-channel message transmission in a mutually coupled semiconductor lasers system. *Optics Communications*, 282(11), pp. 2217-2222, 2009.

[57] M.P. Kennedy. Robust Op-Amp realization of Chua's circuit. *Frequenz*, 46, pp. 66-80, 1992.

[58] M.P. Kennedy. Chaos in the colpitts oscillator. *IEEE Trans. Circuits & Syst.-II*, 41, pp. 771-774, 1994.

[59] R. Kılıç. A comparative study on realizations of Chua's circuit: Hybrid realizations of Chua's circuit combining the circuit topologies proposed for Chua's diode and inductor elements. *Int. Journal of Bifurcation and Chaos*, 13, pp. 1475-1493, 2003.

[60] R. Kılıç. On CFOA-based realizations of Chua's circuit. *Circuits, Systems and Signal Processing*, 22, pp. 475-491, 2003.

[61] R. Kılıç. Experimental study on inductorless CFOA-based Chua's circuit. *Int. Journal of Bifurcation and Chaos*, 14, pp. 1369-1374, 2004.

[62] R. Kılıç. A harmony of linear and nonlinear oscillations: Wien Bridge-based mixed-mode chaotic circuit. *Journal of Circuits, Systems and Computers*, 13, pp. 137-149, 2004.

[63] R. Kılıç. Chaos synchronization in SC-CNN-based circuit and an interesting investigation: Can a SC-CNN-based circuit behave synchronously with the original Chua's circuit. *Int. Journal of Bifurcation and Chaos*, 14, pp. 1071-1083, 2004.

[64] R. Kılıç. Impulsive synchronization between two mixed-mode chaotic circuit. *Journal of Circuits, Systems and Computers*, 14, pp. 333-346, 2005.

[65] R. Kılıç. Experimental modifications of VOA-based autonomous and nonautonomous Chua's circuits for higher dimensional operation. *Int. Journal of Bifurcation and Chaos*, 16(9), pp. 2649-2658, 2006.

[66] R. Kılıç. Experimental investigation of impulsive synchronization between two mixed-mode chaotic circuit. *Int. Journal of Bifurcation and Chaos*, 16(5), pp. 1527-1536, 2006.

[67] R. Kılıç. Mixed-mode chaotic circuit with Wien-Bridge configuration: The Results of experimental verification. *Chaos, Solitons and Fractals*, 32, pp. 1188-1193, 2007.

[68] R. Kılıç and M. Alçı. Design and analysis of the mixed-mode chaotic circuit. *COC'2000: 2nd International Conference on Control of Oscillations and Chaos*, St. Petersburg, Russia, vol. 1, pp. 158-159, 2000.

[69] R. Kılıç and M. Alçı. Mixed-mode chaotic circuit and its application to chaotic communication for transmission of analog signals. *Int. Journal of Bifurcation and Chaos*, 11, pp. 571-581, 2001.

[70] R. Kılıç and M. Alçı. Chaotic switching system using mixed-mode chaotic circuit. *Proc. of IEEE MIDWEST symposium on Circuit and Systems*, Dayton, Ohio, USA, vol. 2, pp. 584-587, 2001.

[71] R. Kılıç, M. Alçı and E. Günay. A SC-CNN-based chaotic masking system with feedback, *Int. Journal of Bifurcation and Chaos*, 14, pp. 245-256, 2004.

[72] R. Kılıç, M. Alçı and E. Günay. Two impulsive synchronization studies using SC-CNN-based circuit and Chua's circuit. *Int. Journal of Bifurcation and Chaos*, 14, pp. 3277-3293, 2004.

[73] R. Kılıç, M. Alçı and M. Tokmakçı. Mixed-mode chaotic circuit. *Electronics Letters*, 36, pp. 103-104, 2000.

[74] R. Kılıç, U. Çam, M. Alçı and H. Kuntman. Improved realization of mixed-mode chaotic circuit. *Int. Journal of Bifurcation and Chaos*, 12(6), pp. 1429-1435, 2002.

[75] R. Kılıç, U. Çam, M. Alçı, H. Kuntman and E. Uzunhisarıklı. Realization of Chua's circuit using FTFN-based nonlinear resistor and inductance simulator. *Frequenz*, 58(1-2), pp. 37-41, 2004.

[76] R. Kılıç and B. Karauz. Implementation of a laboratory tool for studying chaotic MMCC circuit, *Int. Journal of Bifurcation and Chaos*, 17(10), pp. 3633-3638, 2007.

[77] R. Kılıç and B. Karauz. Chaos training boards: Versatile pedagogical tools for teaching chaotic circuits and systems. *The International Journal of Engineering Education*, 24(3), pp. 634-644, 2008.

[78] R. Kılıç. Ö.G. Saracoğlu and F. Yıldırım. A New Nonautonomous Version of Chua's Circuit: Experimental Observations. *Journal of the Franklin Institute*, 343, (2), pp. 191-203, 2006.

[79] R. Kılıç and F. Yıldırım. Wien Bridge-based mixed-mode chaotic circuit (W-MMCC): An educational chaotic circuit model for studying linear and nonlinear oscillations. *The 11th International Conference on Nonlinear Dynamics of Electronics Systems (NDES)*, Scuol, Switzerland, 121-125, 2003.

[80] R. Kılıç and F. Yıldırım. Current-Feedback Operational Amplfier-based inductorless mixed-mode Chua's circuits. *Int. Journal of Bifurcation and Chaos*, 16, pp. 709-714, 2006.

[81] R. Kılıç and F. Yıldırım Dalkıran. FPAA-based programmable implementation of a chaotic system characterized with different nonlinear functions. *2008 International Symposium on Nonlinear Theory and its applications NOLTA*, Budapest, Hungary, 132-138, 2008.

[82] R.Kılıç and F. Yıldırım Dalkıran. A survey of Wien bridge-based chaotic oscillators: Design and experimental issues. *Chaos, Solitons and Fractals*, 38(5), pp. 1394-1410, 2008.

[83] R. Kılıç and F. Yıldırım Dalkıran. Utilizing SIMULINK in modeling and simulation of generalized chaotic systems with multiple nonlinear functions. *Computer Applications in Engineering Education*, DOI: 10.1002/cae.20273, 2009.

[84] R. Kılıç and F. Yıldırım Dalkıran. Reconfigurable implementations of Chua's circuit. *Int. Journal of Bifurcation and Chaos*, 19(4), pp. 1339-1350, 2009.

[85] L. Kocarev, K.S. Halle, K. Eckert and L.O. Chua. Experimental demonstration of secure communication via chaotic synchronization. *Int. Journal of Bifurcation and Chaos*, 2(3), pp. 709-713, 1992.

[86] M. Lakshmanan and K.Murali, Chaos in nonlinear oscillators: Controlling and synchronization, *World Scientific*, 1996.

[87] X. Li, K.E. Chlouverakis and D.Xu. Nonlinear dynamics and circuit realization of a new chaotic flow: A variant of Lorenz, Chen and Lü. *Nonlinear Analysis: Real World Applications*, 10(4), pp. 2357-2368, 2009.

[88] K-Y. Lian, P. Liu, T-S. Chiang and C-S. Chiu. Adaptive synchronization design for chaotic systems via a scalar driving signal. *IEEE Trans. on Circuits & Syst.-I*, 49(1), pp. 17-27, 2002.

[89] E. Lindberg, K. Murali and A. Tamasevicius. The smallest transistor-based nonautonomous chaotic circuit. *IEEE Trans. on Circuits and Syst.-II*, 52(10), pp. 661-664, 2005.

[90] K.E. Lonngren, Notes to accompany a student laboratory experiment on chaos, *IEEE Trans. on Education*, 34, pp. 123-128, 1991.

[91] G. Manganaro, P. Arena and L. Fortuna. Cellular neural networks: Chaos, complexity and VLSI processing, *Springer-Verlag*, 1999.

[92] T. Matsumoto. A chaotic attractor from Chua's circuit. *IEEE Trans. on Circuits & Syst.*, CAS-31(12), pp. 1055-1058, 1984.

[93] T. Matsumoto, L.O. Chua and M. Komuro. The double scroll bifurcations. *Int. Journal Circuit Theory and Appl.*, 14(1), pp. 117-146, 1986.

[94] T. Matsumoto, L.O. Chua and M. Komuro. The double scroll. *IEEE Trans. on Circuits & Syst.*, CAS-32(8), pp. 797-818, 1985.

[95] V. Milanovic and M.E. Zaghloul. Improved masking algorithm for chaotic communications systems. *Electronics Letters*, 32(1), pp. 11-12, 1996.

[96] L. Min and N. Yu. Some analytical criteria for local activity of two-port CNN with three or four state variables: analysis and applications. *Int. Journal of Bifurcation and Chaos*, 12(5), pp. 931-963, 2002.

[97] Ö. Morgül. An RC realization of Chua's circuit family. *IEEE Trans. Circuits & Syst.-I*, 47(9), pp. 1424-1430, 2000.

[98] M. Mulukutla and C. Aissi. Implementation of the Chua's circuit and its applications, *Proc. of the 2002 ASEE Gulf-Southwest Annual Conference*, Lafayette, USA, 2002.

[99] A.P. Munuzuri, J.A.K. Suykens and L.O. Chua. A CNN approach to brain-like chaos-periodicity transitions. *Int. Journal of Bifurcation and Chaos*, 8(11), pp. 2263-2278, 1998.

[100] K. Murali and M. Lakshmanan. Synchronizing chaos in driven Chua's circuit. *Int. Journal of Bifurcation and Chaos*, 3(4), pp. 1057-1066, 1993.

[101] K. Murali, M. Lakshmanan and L.O. Chua. Bifurcation and chaos in the simplest dissipative non-autonomous circuit. *Int. Journal of Bifurcation and Chaos*, 4(6), pp. 1511-1524, 1994.

[102] S. Nakagawa and T. Saito. An RC OTA hysteresis chaos generator. *IEEE Trans. on Circuits & Syst-I*, 43(12), pp. 1019-1021, 1996.

[103] A. Namajunas and A. Tamasevicius. Modified Wien-bridge oscillator for chaos. *Electronic Letters*, 31(5), pp. 335-336, 1995.

[104] A. Namajunas and A. Tamasevicius. Simple RC chaotic oscillator. *Electronic Letters*, 32(11), pp. 945-946, 1996.

[105] N.A. Natsheh, J.G. Kettleborough and J.M. Nazzal. Analysis, simulation and experimental study of chaotic behaviour in parallel-connected DC–DC boost converters. *Chaos, Solitons & Fractals*, 39(5), pp. 2465-2476, 2009

[106] H.H. Nien, C.K. Huang, S.K. Changchien, H.W. Shieh, C.T. Chen and Y.Y. Tuan. Digital color image encoding and decoding using a novel chaotic random generator. *Chaos, Solitons & Fractals*, 32(3), pp. 1070-1080, 2007.

[107] K. O'Donoghue, M.P. Kennedy, P. Forbes, M. Qu and S. Jones. A fast and simple implementation of Chua's oscillator with Cubic-like nonlinearity. *International Journal of Bifurcation and Chaos,* 15, pp. 2959-2971, 2005.

[108] M.J. Ogorzalek. Taming chaos-I: Synchronization. *IEEE Trans. on Circuits & Syst.-I*, 40(10), pp. 693-699, 1993.

[109] M.J. Ogorzalek. Chaos and complexity in nonlinear electronic circuits. *World Scientific*, 1997.

[110] A. Oksasoglu and T. Akgül. Chaotic masking scheme with a linear inverse system. *Physical Review Letters*, 75(25), pp. 4595-4597, 1995.

[111] A.I. Panas, T. Yang and L.O. Chua. Experimental results of impulsive synchronization between two Chua's circuits. *Int. Journal of Bifurcation and Chaos*, 8(3), pp. 639-644, 1998.

[112] L.M. Pecora and T.L. Carroll. Synchronization in chaotic systems. *Phys. Rev. Letters*, 64, pp. 821-824, 1990.

[113] V. Perez-Munuzuri, V. Perez-Villar and L.O. Chua. Autowaves for image processing on a two-dimensional CNN array of excitable nonlinear circuits: flat and wrinkled labyrinths, *IEEE Trans. on Circuits & Syst.-I*, 40(3), pp. 174-181, 1993.

[114] Pico Technology Limited, United Kingdom.

[115] J.Q. Pinkney, P.L. Camwell and R.Davies. Chaos shift keying communications system using self-synchronising Chua oscillators, *Electronics Letters*, 31(13), pp. 1021-1022, 1995.

[116] J.R.C. Piqueira. Using bifurcations in the determination of lock-in ranges for third-order phase-locked loops. *Communications in Nonlinear Science and Numerical Simulation*, 14(5), pp. 2328-2335, 2009.

[117] L. Pivka. Autowaves and spatio-temporal chaos in CNNs-I: A tutorial. *IEEE Trans. on Circuits & Syst.-I*, 42(10), pp. 638-649, 1995.

[118] PSPICE, *Cadence Design Systems. Inc.*

[119] G. Qi, M.A. Van Wyk, B.J. Van Wyk and G. Chen. A new hyperchaotic system and its circuit implementation. *Chaos, Solitons & Fractals*, 40(5), pp. 2544-2549, 2009.

[120] A.G. Radwan, A.M. Soliman and A-L. El-Sedeek. An inductorless CMOS realization of Chua's circuit, *Chaos, Solitons & Fractals*, 18(1), pp. 149-158, 2003.

[121] A.G. Radwan, A.M. Soliman and A. El-Sedeek. MOS realization of the modified Lorenz chaotic system. *Chaos, Solitons & Fractals*, 21(3), pp. 553-561, 2004.

[122] A. Rodriguez-Vazquez and M. Delgado-Restituto. CMOS design of chaotic oscillators using variables: A Monolithic Chua's circuit. *IEEE Trans. on Circuits & Syst.-II*, 40(10), pp. 596-611, 1993.

[123] T. Roska and L.O. Chua. The CNN universal machine: An analogicarray computer. *IEEE Trans. on Circuits & Syst.-I*, 40(3), pp. 163-173, 1993.

[124] M. Rostami, S.H. Fathi, M. Abedi and C. Lucas. Switch time bifurcation elimination analysis in SVC plants. *Electric Power Systems Research*, 74(2), pp. 177-185, 2005.

[125] P.K. Roy, S. Chakraborty and S.K. Dana. Experimental observation on the effect of coupling on different synchronization phenomena in coupled nonidentical Chua's oscillators. *Chaos*, 13(1), pp. 342-355, 2003.

[126] E. Sanchez, M.A. Matias and V. Perez-Munuzuri. Chaotic synchronization in small assemblies of driven Chua's circuits. *IEEE Trans. on Circuits & Syst.-I*, 47(5), pp. 644-654, 2000.

[127] F.A. Savacı and J. Vandewalle. On the stability analysis of cellular neural networks. *IEEE Trans. on Circuits & Syst.-I*, 40(3), pp. 213-215, 1993.

[128] R. Senani and S.S. Gupta. Implementation of Chua's chaotic circuit using current feedback op-amps. *Electronic Letters*, 34(9), pp. 829-830, 1998.

[129] Z. Shi, S. Hong and K. Chen. Experimental study on tracking the state of analog Chua's circuit with particle filter for chaos synchronization. *Physics Letters A*, 372(34), pp. 5575-5580, 2008.

[130] SIMULINK/MATLAB, *The MathWorks, Inc.*

[131] J.C. Sprott. A new class of chaotic circuit. *Physics Letters A*, 266, pp. 19-23, 2000.

[132] J.C. Sprott. Simple chaotic systems and circuits. *Am. J. Phys*, 68(8), pp. 758-763, 2000.

[133] J.A.K. Suykens and L.O. Chua. N-double scroll hypercubes in 1-D CNNs. *Int. Journal of Bifurcation and Chaos*, 7(8), pp. 1873-1885, 1997.

[134] J.A.K. Suykens, P.F. Curran and L.O. Chua. Master-slave synchronization using dynamic output feedback. *Int. Journal of Bifurcation and Chaos*, 7(3), pp. 671-679, 1997.

[135] T. Suzuki and T. Saito. On fundamental bifurcations from a hysteresis hyperchaos generator. *IEEE Trans. on Circuits & Syst.-I*, 41(12), pp. 876-884, 1994.

[136] E. Swiercz, A new method of detection of coded signals in additive chaos on the example of Barker code, *Signal Processing*, 86(1), pp. 153-170, 2006.

[137] A. Tamasevicious, G. Mykolaitis, V. Pyragas and K. Pyragas. A simple chaotic oscillator for educational purposes. *Eur. J. Phys*. 26, pp. 61-63, 2005.

[138] K.S. Tang, K.F. Man, G.Q. Zhong and G. Chen. Generating chaos via x|x|. *IEEE Trans. on Circuits & Syst.-I*, 48, pp. 635-641, 2001.

[139] L.A.B. Torres and L.A. Aguirre. Inductorless Chua's circuit. *Electronic Letters*, 36(23), pp. 1915-1916, 2000.

[140] J.E. Varrientos and E. Sanchez-Sinencio. A 4-D chaotic oscillator based on a differential hysteresis comparator. *IEEE Trans. on Circuits & Syst.-I*, 45(1), pp. 3-10, 1998.

[141] R. Vázquez-Medina, A. Díaz-Méndez, J.L. Del Río-Correa and J. López-Hernández. Design of chaotic analog noise generators with logistic map and MOS QT circuits. *Chaos, Solitons & Fractals*, 40(4), pp. 1779-1793, 2009.

[142] U.E. Vincent, A. Ucar, J.A. Laoye and S.O. Kareem. Control and synchronization of chaos in RCL-shunted Josephson junction using backstepping design. *Physica C: Superconductivity*, 468(5), pp. 374-382, 2008.

[143] M.E. Yalçın, J.A.K. Suykens and J.P.L. Vandewalle. Cellular Neural Networks, Multi-Scroll Chaos and Synchronization. *World Scientific*, 2005.

[144] T. Yang and L.O. Chua. Secure communication via chaotic parameter modulation. *IEEE Trans. on Circuits & Syst.-I*, 43(9), pp. 817-819, 1996.

[145] T. Yang and L.O. Chua. Impulsive control and synchronization of nonlinear dynamical systems and application to secure communication. *Int. Jounal Bifurcation and Chaos*, 7(3), pp. 645-664, 1997.

[146] X-S. Yang and Q. Li. Chaos generator via Wien-bridge oscillator. *Electronics Letters*, 38(13), pp. 623-625, 2002.

[147] L. Yang and T. Yang. Synchronization of Chua's circuits with parameter mismatching using adaptive model-following control. *Chinese Journal of Electronics*, 6(1), pp. 90-96, 1997.

[148] Y-Z. Yin. Experimental demonstration of chaotic synchronization in the modified Chua's oscillators. *Int. Jounal Bifurcation and Chaos,* 7(6), pp. 1401-1410, 1997.

[149] S. Yu, J. Lü and G. Chen. A family of *n*-scroll hyperchaotic attractors and their realization. *Physics Letters A*, 364(3-4), pp. 244-251, 2007.

[150] G.Q. Zhong. Implementation of Chua's circuit with a cubic nonlinearity. *IEEE Trans. on Circuits & Syst.-I*, 41, pp. 934-941, 1994.

[151] H. Zhu, S. Zhou and J. Zhang. Chaos and synchronization of the fractional-order Chua's system, *Chaos, Solitons & Fractals*, 39(4), pp. 1595-1603, 2009.

Index

.